Ian,

Much Love

Karel xxx

Xmas '92

D1426724

# SCIENCE
*with a*
*Smile*

# SCIENCE

*with a*
*Smile*

An anthology selected by
Robert L Weber

Institute of Physics Publishing
Bristol and Philadelphia

*British Library Cataloguing-in-Publication Data*
A catalogue record for this book is available from the British Library.

ISBN 0-7503-0211-9


*Library of Congress Cataloging-in-Publication Data are available*


IOP Publishing Ltd has attempted to trace the copyright holder of all the material reproduced in this publication and apologizes to copyright holders if permission to publish in this form has not been obtained.


Published by IOP Publishing Ltd, a company wholly owned by the Institute of Physics, London

IOP Publishing Ltd
Techno House, Redcliffe Way, Bristol BS1 6NX, England

US Editorial Office: IOP Publishing Inc., The Public Ledger Building, Suite 1035, Independence Square, Philadelphia, PA 19106

Typeset by TVB Bookwork, Salisbury, Wiltshire
Printed in the UK by Cambridge University Press

For my grandchildren
Alexandra, James and Katja

# Preface: Science with a Smile

Academicians tend to analyse and theorize. Humor seems to be surprisingly refractory to both processes. During the Renaissance, writers who examined humor were concerned mainly with the corrective function of comedy. With Descartes we find for the first time emphasis upon psychological rather than literary or ethical aspects of humor. But down to the present, writers who have analysed humor have presented either so limited a view or such diffuse, unsupported conclusions as to make it inappropriate to designate the work as a *theory* of humor. Rather we might speak of concepts: the superiority concept of humor, the incongruity concept, the release-from-constraint concept, the ambivalence concept, etc.

*Science with a Smile* attempts to offer no theory of humor. This anthology is intended for enjoyment. But outlasting one's initial smile may often be some appreciation of self-criticism, changing attitude in science, and illumination of personalities important to science.

This anthology of humorous history of science is intended for casual, intermittent reading. Its goal is not analysis but rather enjoyment. It may lead you to a warm feeling toward individual scientists and respect for the ingenuity and cheerfulness of the scientific spirit.

As a graduate student studying physics, I asked several professors to recommend reading in the history of science. I was referred to George Sarton's *History of Science*, but with the unmistakable hint that doing science was more important than reading history. This was sound advice especially for one just starting a career in an experimental science. But in recent years I have found satisfaction in that several prestigious universities have endowed professorships in the history of science. I think they vindicate my early feeling that there is something worthy about that concern with science and its importance to society.

When I first developed an interest in humor in physics I had difficulty in locating examples, beyond ephemeral examination boners. Humor was rarely indexed; some was unintentional. An MIT professor gave a cordial answer to my inquiry about a particular item but added that in his opinion I was engaging in useless research. As my correspondence gradually expanded worldwide, I found kindred spirits and many gems of humor. Recently several scholarly societies concerned with the serious study of science humor have been formed. Some journals of science and technology have long had a humor column or an April 1 spoof. But journals especially devoted to science humor have not faired well. Under relentless pressure of meeting deadlines, editors seem to allow humor to degenerate into crude sexual exposures. Free from such pressure was the *Journal of Jocular Physics* published at five-year intervals to celebrate Niels Bohr's birthday.

I wish to express gratitude to the staff of the Books Department of IOP Publishing, Ltd, for crucial help and friendly cooperation in the publishing of this volume and two earlier anthologies. Special thanks to Dr Dorothy G Fisher, Neville Hankins, Martin J S Beavis, Sean Pidgeon, Maureen Clarke, Graham Douglas, Al Troyano and Jenny Pickles.

The Institute of Physics and I would like to thank all those who provided articles and gave permission to publish them in this book.

**Robert L Weber**

# Contents

# Contents

Contents

# Personalities

Every honest researcher I know admits he's just a professional amateur. He's doing whatever he's doing for the first time. That makes him an amateur. He has enough sense to know that he's going to have a lot of trouble, so that makes him a professional.

*Charles F Kettering*


A modern poet has characterized the personality of art and the impersonality of science as follows: Art is I; Science is We.

*Claude Bernard*

# Astronomical poetry and quips

Contributed by
Louis Berman

What was God doing before He made Heaven and Earth?
He was creating Hell for people who ask questions like
that.

*St Augustine*

There is nothing more incomprehensible than a wrangle
among astronomers.

*Henry L Mencken*

From *Space and
Time in the
Modern
Universe*, p 211

Ten years of radio astronomy have taught humanity more
about the creation and organization of the universe than
thousands of years of religion and philosophy.

*P C W Davis*

From *Essay on
Nature*

The astronomers said: 'Give us matter, and a little motion,
and we will construct the universe.'

*Ralph Waldo Emerson*

*Big whirls have little whirls
That feed on their velocity
And little whirls have lesser whirls
And so on to viscosity.*

*F L Richardson*

**Some sort of game**

*John Ciardi*

*Toy-maker Ptolemy
Made up a universe
Nine crystal yo-yos he
Spun on one string. It was
Something to see it go,
Half sad to see it pass.*

## When I heard the learn'd astronomer

*Walt Whitman*

*When I heard the learn'd astronomer,*
*When the proofs, the figures, were ranged in columns before*
   *me,*
*When I was shown the charts and diagrams,*
   *to add, divide, and measure them,*
*When I sitting heard the astronomer when he*
   *lectures with much applause in the lecture-room,*
*How soon unaccountably I became tired and sick,*
*Till rising and gliding out I wander'd off by myself,*
*In the mystical moist night air, and from time to time,*
*Looked up in perfect silence at the stars.*

Analyzing a spectrum is exactly like doing a crossword puzzle, but when you get through with it, you call the answer research.

*Henry Norris Russell*

---

# Antimatter concern

Question and Answer, contributed by A P French

July 1982

Dear Sirs:

Being a safety minded individual I thought I would write you before experimenting on my own. Is it safe to mix Antipasto and Pasta together and could this be a future energy supply?

Awaiting your reply,

Sincerely,
Nyles Bauer
59B Locust Ave
New Rochelle, NY 10801

MASSACHUSETTS INSTITUTE OF TECHNOLOGY
DEPARTMENT OF PHYSICS
CAMBRIDGE, MASSACHUSETTS 02139
July 16, 1982

Mr. Nyles Bauer
59B Locust Ave.
New Rochelle, NY 10801

Dear Mr. Bauer:

I believe that your thoughtful and interesting suggestion about the mixing of pasta and antipasto deserves some acknowledgment. This process might well be a significant energy source—but only, I think, intragastrically. I estimate that the digestion of 1 lb of the mixture would release energy equivalent to about 0.001 megawatt hours or 0.000001 kilotons of TNT. I would not foresee any unusual hazards.

Sincerely yours,

A P French
Professor of Physics
Academic Officer, Physics Dept.

---

# New genesis

*George Gamow*

In the beginning God created radiation and ylem. And ylem was without shape or number, and the nucleons were rushing madly over the face of the deep.

And God said: 'Let there be mass two'. And there was mass two. And God saw deuterium, and it was good.

And God said: 'Let there be mass three'. And there was mass three. And God saw tritium and tralphium, and they were good. And God continued to call number after number until He came to transuranium elements. But

16

when He looked back on his work He found that it was not good. In the excitement of counting, He missed calling for mass five and so, naturally, no heavier elements could have been formed.

God was very much disappointed, and wanted first to contract the Universe again, and to start all over from the beginning. But it would be much too simple. Thus, being almightly, God decided to correct his mistake in a most impossible way.

And God said: 'Let there by Hoyle'. And there was Hoyle. And God looked at Hoyle ... and told him to make heavy elements in any way he pleased.

And Hoyle decided to make heavy elements in stars, and to spread them around by supernovae explosions. But in so doing he had to obtain the same abundance curve which would have resulted from nucleosynthesis in ylem, if God would not have forgotten to call for mass five.

And so, with the help of God, Hoyle made heavy elements in this way, but it was so complicated that nowadays neither Hoyle, nor God, nor anybody else can figure out exactly how it was done.

Amen.

(My attitude toward the steady-state theory, expressed in this piece, may account for my not receiving an invitation to the 1958 Solvay Congress on cosmology.)

---

Louis Phillips
*Academe*
February 1979
p 48

## Socrates among the Athenians

# An interview with Charles Babbage

Alex Ragen
*Datamation* **23**
April 1977 p 100

[This transcript originated from an as yet unidentified node of the ARPANET.]

*Question:*   As the man who, so to speak, set the ball rolling back in 1812, what do you think of the state of computer science in 1977?

*Babbage:*   I'm rather pleased by it all, I'd say. Of course, I've always had the greatest confidence in the future of difference engines, or as you'd call them today, computers. In my time the idea failed to catch on for a number of reasons—incidentally, I don't agree with the so-called historical experts about why I never succeeded in constructing a working model. Perhaps we'll discuss that point later on.

As I said, I'm rather pleased by it all but I'm afraid I can't say the same for my friend George Boole. He's been in a terribly depressed state these last thirty years or so. You see, it was always one of the great points of pride in his life that his most important work—his Boolean algebra—was an utterly useless collection of aesthetic niceties which no engineer could ever put to practical use in bridge building or wheel grinding or any of the dirty little projects engineers are always dreaming up. Instead he sees now these giant collections of nuts and bolts and transistors and resistors and IC's and LED's and RAML's and PROM's and the devil knows what else, lights flashing, wheels spinning, noise, *et cetera*—everything the poor chap has always detested and he knows that none of it would have been possible without his own most treasured invention—this algebra of his. Do you realize that he hasn't said a word to Von Nuemann since he arrived?

*Question:*   How is Lady Lovelace getting on?

*Babbage:*   Ah, dear old Ada. An absolutely exquisite woman—as charming here as she ever was down there. She's doing quite well. You know, she's been reading (we have regular subscriptions to all the periodicals here) that history now regards her as the world's first programmer. Quite a distinction, I daresay, for a member of the gentler

18

sex, although I understand that's been changing as well. She's quite flattered by all the attention she's been receiving.

Still, there is a batch of grey on her horizon. Her relations with her father are strained. He had always wanted her to follow in his footsteps and become a woman of letters, but she sided with her mother instead. You see, Byron never did well at school and especially despised mathematics, whereas Lady Byron was a splendid amateur mathematician. 'Princess Parallelogram' he used to call her. What a disappointment it was for him that their daughter chose mathematics over poetry. You must realize that she was an only child. He was so heartbroken when he discovered that her mother had taught her seven proofs of the Pythagorean Theorem by the time the child was five that he ran off to Greece to drive out the infidel Turks, or something like that. What lives they all had.

*Question:* Do you see Herman Hollerith from time to time?

*Babbage:* Not anymore, really. These Americans just don't seem to fit in here in our little group. Nevertheless, I believe he might have managed to adjust socially had he only been more open minded.

*Question:* In what way?

*Babbage:* You see, he was so involved with his invention—those punched card machines—that he never got round to seeing the subject of difference engines, that is, computers, from any other point of view. He has a tendency to regard everything as an extension of those little cards, and I'm afraid he hasn't been able to keep pace with the developments of the last fifteen or twenty years. He's become anchored in one way of thinking. What a pity! He's a bright fellow and quite an inventor, but his time has passed. Even Jacquard can't tolerate him.

*Question:* You mentioned Dr Von Neuman before. Can you tell us something about him?

*Babbage:* Well, there really isn't much to tell. You see, he is one of those chaps who takes his work altogether too seriously and it appears that once he finished with the process of moving in here, instead of relaxing and enjoying his well deserved holiday, he's gone right back to work. He regards it all as a kind of game, he once told me. 'Lots of fun,' I believe he said. Well, I say that he takes games too seriously for his own good. And since Boole won't have anything to do with him, he's being ostracized socially. Too bad—I understand he has an excellent sense of humor.

*Question:* Turning from the subject of personalities to that of technology, in what respects do you think the world has changed as a result of computers and to what extent has this, in your opinion, been a positive development?

*Babbage:* Well, I would say that the most important result of computerization—I believe that is the word most commonly used to describe the process—is that life has lost its leisurely pace. There was a time when people carried watches about with them for no other reason than to identify themselves as members of a class that was wealthy enough to afford watches. Nobody cared a tuppence for the time of day. There were no typewriters or copying machines, the posts were slow, even gossip took days to travel from one village to another and no one was in much of a hurry about anything. Today it seems that humanity can't wait a nanosecond to invert a matrix.

I think as well that altogether too much information is available to people these days. I don't mean the kind of things that people need to know but before computers weren't able to know, but rather the kind of useless information that no one really wants but that everyone feels obligated to demand because the capability for obtaining this unnecessary information exists. Then someone establishes an international communications network so that thousands of people all over the world can have this nonsense at their fingertips. All of this is very far removed from what I had in mind when I developed the difference engine.

*Question:* Exactly what did you have in mind? How did you imagine that your difference engines would change the world?

*Babbage:* To be perfectly frank, my dear fellow, I was remarkably

20

naïve about the possibilities. I assumed that it would be of no use to anyone but mathematicians. In those days, you know, solving a system of 500 linearly independent simultaneous equations in 500 unknowns was a life's work, not to mention the job of determining linear independence. Some poor chap would spend a few years at Oxford learning mathematics to tackle the job and then shut himself up in a room somewhere for thirty years to work on the problem. Today a difference engine, or I should say a computer, finishes the job in an hour or so. Now that's the kind of thing I had in mind—making life tolerable for mathematics Dons and giving them an opportunity to marry.

I never imagined that a difference engine would make such a difference. I'm astounded at the revolution that's taken place.

As for the size of the newer models, I'm quite at a loss for words. My original design was smaller than the ENIAC, as you may be aware, and I always considered that machine to be inferior because of its unnecessary bulk. The microprocessors which are so popular now are clearly a step in the right direction.

Incidentally, a development which gives me great personal pain is the shameless exploitation of my distinguished family name on the part of unscrupulous persons who consider the possibility of gaining a few shillings at the expense of those who cannot defend themselves an acceptable pursuit. I must take this opportunity to register my disapproval.

*Question:* You mentioned earlier that historians have an inaccurate understanding of why you never succeeded in constructing a working model of your difference engine. I wonder if you might care to enlighten us about the real reasons for your lack of success.

*Babbage:* Well, there were a number of problems that I was unable to overcome. You must understand that my idea was many decades ahead of its time and there were some strong vested interests who were determined that I should fail. There was for example a conspiracy among actuaries and statisticians who were afraid they might lose their positions if I succeeded. In Oxford a number of distinguished professors who hadn't the faintest conception of what I was doing went into a panic when they heard from some stupid Member of Parliament that I was working on a machine that would make professors of mathematics obsolete. Sheer nonsense of course, but those senile old gentlemen had a great deal of influence and very nearly managed to have my research funds cut off.

I had what you would call today software problems, but I don't believe that they were insurmountable. My most serious problem was quite interesting—whenever I was halfway through the construction of one part of the engine, I would have a brilliant idea about how I could rebuild it from scratch and make significant improvements. I would then tear weeks of hard work to pieces and start all over again. That's how I came to my analytical engine, by the way. Of course I made very slow progress this way.

*Question:*     Since you have access to privileged information, Mr Babbage, perhaps you'd care to tell us something about what the future holds for computer science.

*Babbage:*     Well, I don't really know about the future of computer science, but I can give you a few hints about the future of computer scientists. You know, I haven't exactly been idle these last few years. Did you notice that there aren't any long queues at the pearly gates anymore? Would you care to know why? St Peter has a terminal now, hooked up to a huge data base, to help him make those on-the-spot decisions. We don't miss a trick anymore, my dear chap, so watch your step.

---

# Blake and fractals

J D Memory
*Mathematics Magazine*
63.4 October 1990
p 280

*William Blake said he could see*
*Vistas of infinity*
*In the smallest speck of sand*
*Held in the hollow of his hand.*

*Models for this claim we've got*
*In the work of Mandelbrot:*
*Fractal diagrams partake*
*Of the essence sensed by Blake.*

*Basic forms will still prevail*
*Independent of the scale;*
*Viewed from far or viewed from near*
*Special signatures are clear.*

*When you magnify a spot,*
*What you had before, you've got.*
*Smaller, smaller, smaller yet,*
*Still the same details are set;*

*Finer than the finest hair*
*Blake's infinity is there,*
*Rich in structure all the way—*
*Just as the mystic poets say.*

## Walter Nernst

Edgar W Kutzscher, in a letter.

[Professor Nernst liked to 'spice' his lectures with many interesting stories, anecdotes, and jokes. We, as students, appreciated this habit tremendously and I remember that even his first assistant, when he had to take over the lecture in the absence of Professor Nernst, used to say, 'At this particular point, Professor Nernst likes to tell such-and-such a story.' One story I remember went something like this:]

Nernst, after discussing the first and second fundamental laws of thermodynamics, used to say, 'As you remember, the *first* fundamental law of thermodynamics was formulated and theoretically proved by *three* physicists, Robert Mayer, James Prescott Joule, and Herman von Helmholtz—the second law of thermodynamics was formulated by *two* great physicists, Rudolph Clausius and William Thompson—and I developed the *third* law alone. So you see, first law by three, second law by two, third law by one—which proves a fourth fundamental law of thermodynamics cannot exist.'

23

# Peter Debye

Contributed by James E Sturm, attributed to W Lipscomb

Peter Debye, professor at Cornell University in the 1940s, frequently took the Lehigh Valley Railroad to get to such cities as Philadelphia or Buffalo. On one occasion, when the conductor came by for tickets, Debye was hurriedly fumbling in his pockets to find his ticket. The conductor said, 'Don't worry about it; I'll pick it up when you get off'. Debye, still non-plussed, replied, 'But I don't know where I'm going!'

---

1963 *Applied Physics* **2** 973

The man of science appears to be the only man who has something to say, just now—and the only man who does not know how to say it.

Sir James M Barrie

---

# Systems analysis

Apocryphal anecdote communicated by Amron H Katz

President Johnson called in Secretary of Defense McNamara one day, during the Vietnam war, and asked him 'How are we doing, Mac?' Secretary McNamara stepped over to a large sketch pad and printed [the capitalized part], 'Mr President you are faced with an

INEFFICIENT EXPENSIVE LOSS'

L B J promptly asked 'What are you going to do about that?' McNamara responded, 'Mr President, we have some bright systems analysts in my office, at the RAND Corp., at IDA [the Institute of Defense Analysis] and elsewhere. We are going to convert the inefficient expensive loss into [and he printed just below his first entry on the pad] an

EFFICIENT CHEAP LOSS'

The president hesitated for a few milliseconds and asked 'Why the hell don't you figure out how to win?' To which McNamara said, 'I'm sorry, Mr President, going from LOSS to WIN requires an invention. Systems Analysis works only on the adjectives!'

# Einstein said

Everything should be made as simple as possible but not simpler.

Nationalism is an infantile disease; it is the measles of mankind.

No path leads from a knowledge of that which is to that which should be.

If you are out to describe the truth leave elegance to the tailor.

The World War after the next one will be fought with rocks.

I think and think for months, for years; 99 times the conclusion is false, but the hundredth time I am right.

Sit with a pretty girl for an hour, and it seems like a minute; sit on a hot stove for a minute, and it seems like an hour—that's relativity.

On the statistical method of quantum mechanics: I cannot believe that God plays dice with the world.

My religion consists of a humble admiration of the illimitable superior spirit who reveals himself in the slight details we are able to perceive with our frail and feeble minds. That deeply emotional conviction of the presence of a superior reasoning power, which is revealed in the incomprehensible universe, forms my idea of God.

Science without religion is lame, religion without science is blind.

I believe in Spinoza's God who reveals Himself in the orderly harmony of what exists, not in a God who concerns himself with the fates and actions of human beings.

In *Albert Einstein, Creator and Rebel* by Banesh Hoffman 1972 (New York: Viking) p 24

To punish me for my contempt for authority, Fate made me an authority myself.

# Einstein said

Inscribed at
Princeton

God is cunning but He is not malicious. [*Raffiniert ist der Herr Gott aber boschaft ist Er nicht.*]

In *Albert
Einstein: A
Documentary
Biography* by C
Seelig 1956
(London: Staples)

I can, if the worst comes to the worst, still realize that the Good Lord may have created a world in which there are no natural laws. In short, chaos. But that there should be statistical laws with definite solutions, i.e., laws which compel the Good Lord to throw the dice in each individual case, I find highly disagreeable.

In *Quest* by L
Infeld 1942
(London:
Gollancz) p 279

God does not care about our mathematical difficulties. He integrates empirically.

'Einstein's
Philosophy of
Science' by
Philipp Frank in
*Review of
Modern Physics*
**21** 349 (1949)

About ten years ago I spoke with Einstein about the astonishing fact that so many ministers of various denominations are strongly interested in the theory of relativity. Einstein said that according to his estimation there are more clergymen interested in relativity than physicists. A little puzzled I asked him how he would explain this strange fact. He answered, a little smiling, 'Because clergymen are interested in the general laws of nature and physicists, very often, are not.' Another day we spoke about a certain physicist who had very little success in his research work. Mostly he attacked problems which offered tremendous difficulties. By most of his colleagues he was not rated very highly. Einstein, however, said about him, 'I admire this type of man. I have little patience with scientists who take a board of wood, look for its thinnest part and drill a great number of holes where the drilling is easy.'

In a 1927 letter to
Max Born, from
*The Born –
Einstein Letters*
by M Born 1971
(New York:
Walker)

What applies to jokes, I suppose, also applies to pictures and to plays. I think they should not smell of a logical scheme, but of a delicious fragment of life, scintillating with various colours according to the position of the beholder. If one wants to get away from this vagueness one must take up mathematics. And even then one

26

reaches one's aim only by becoming completely insubstantial under the dissecting knife of clarity. Living matter and clarity are opposites—they runaway from one another. We are now experiencing this rather tragically in physics.

'Many years ago, when I came to Princeton, I used to go to tea in the math department. There were a bunch of brilliant young chaps there talking mathematics, and I didn't understand a hell of a thing they were saying. I was only a statistician.'

G S Watson

'In a math course you learn whether or not there's a solution; in a physics course you find the solution.'

P A Piroue

'Heresy is the side that loses.'

J V Fleming

# Academic Life

Science is a good piece of furniture for a man to have in an upper chamber provided he has common sense on the ground floor.

*Oliver Wendall Holmes*
1870

# How far can a dog run into the woods?

Harvard University's Richard E Baym takes issue with the answer to an exam question that appeared in this space on March 13:

*Q*: How far can a dog run into the woods?

*A*: Halfway. The rest of the time he is running out.

'The correct answer is "All the way". Certainly we understand that the dog is running "in" only until he reaches the *middle* of the forest, but this is, in fact, all the way in. If the dog ran only halfway "in", he would not yet be at the middle.'

'Indeed if the dog ran halfway in and then ran halfway out, he would still be in the woods.'

It occurs to us that the dog's continued presence there would be useful, in case something happens to that tree that we've been hearing about since high school physics—the one that falls when no one is in the forest and since there is no eardrum to register sound waves, makes no noise.

You know what a fine sense of hearing a dog has. Let him run halfway *in* (or as Mr Baym argues, all the way), settle there, and keep an ear cocked for that tree.

C.G.

---

# Excerpts from *Computer Bluff*

## Introduction

The truth is, none of us could now get by without The Computer. From controlling power stations to working out tax and social security payments. From scheduling jetliners into and out of major international airports—and booking their seats—to analysing medical diagnostic data. Launching moonshots, Saturn spaceprobes, Space Shuttles; forecasting the weather; checking water reservoir levels—and working out home telephone, gas and electricity utility demands. Not to mention printing our salary cheques to enable us to pay them. Twentieth Century, Western industrialised life as we know it today would in just about all aspects be a little more difficult, a little less delightful, occasionally impossible, and in many cases even dangerous or deadly, without The Computer.

And the influence of this vital box of electronic tricks is destined to seep ever more pervasively into every corner of modern life with the spread of a whole new gamut of computer-based 'Information Technology' products. Computer and communications systems incorporating microprocessors, those wizard little chips of silicon intelligence, are increasingly coming to assist us in every conceivable situation: in the factory or office, school and university, home and leisure.

So you had better start bluffing your way in computing. Now. This book is intended to help you do so. It is dedicated to dispelling the myth that computing is for computocratic high priests alone.

On the contrary, computing is creative *fun*. There is already a whole generation of bright-eyed children growing up with their sticky little fingers on computer control keyboards almost, it seems, from the cradle up. *You don't have to be a mathematical genius to appreciate the fundamentals and a few workaday practical principles.*

And if even that seems like too much effort, this book will reveal that there are really only Three Very Important Things to Remember in Computing. Don't put it down, therefore, once you have started reading, until you have mastered at least these three. Then you can get out there and start bluffing. Today.

### BABBAGE

Now Charles Babbage, a Victorian Englishman, is not just a name to remember but the

### First Very Important Thing to Remember (FVITTR)

and therefore, indispensible ammunition in your armoury. Learn it well.

He it was who is credited with the design of the first mechanical computer, in that this non-triumph of mechanical engineering identified and utilised the crucial computer concepts of 'data', 'store', 'process', 'instruction' and, most importantly, 'stored program'. This latter 'stored program' concept means no more nor less than that Babbage perceived the significant breakthrough idea that the *instructions* to tell a computer 'what to do' could be stored *within the computer*, just like the *data* (numbers) upon which the instructions were to operate.

Babbage's first machine, the 'Differencing Engine', could generate mathematical tables of many kinds by the 'method of differences.' But before any model of the 'D.E.' had been built, he became obsessed with a much more revolutionary idea: his 'Analytical Engine'. Unlike any previously designed machine, the 'A.E.' was to possess both a 'store' and a 'mill' (calculating and decision-making unit). These units were to be built of thousands of intricate geared cylinders interlocked in incredibly complex ways. Babbage had a vision of numbers swirling in and out of the mill under control of a program contained in punched cards—an idea inspired by the Jacquard loom, a card-controlled loom that wove amazingly complex patterns.

Babbage, in the best traditions of the dedicated, committed British inventor, died a broken, definitely commercially non-viable man, defeated, it seems by the lack of sufficiently advanced precision engineering expertise which might have turned his metal cog-wheel-ridden monstrosity into working, commercial reality. (Definitely not one of Life's Natural Computer Bluffers, one might cruelly observe.)

Feminists, both male and female, can incidentally take joy in hearing that the name of a woman, **Ada**, Lady Lovelace, figures equally prominently, in its way, in the pioneering saga of Charles Babbage. The Countess of Lovelace (who, would you believe, was Lord Byron's daughter) corresponded with Babbage, contributing, in quite an inspirational way, we think, to his developing theory of the 'stored program'. In an 1842 memoir, she wrote that the 'A.E. might act upon other things besides number.' She suggested that Babbage's Engine, with pitches and harmonics coded into its spinning cylinders, 'might compose elaborate and scientific pieces of music of any degree of complexity or extent.' Finally, she put forward the still definitive 'obedient idiot' view of the programmable computer: 'The Analytical Engine has no pretensions whatever to *originate* anything. It can do whatever we *know how to order it to perform.'*

Thus, as all agree that Babbage's was the first computer, so too do they concur that the bright and beautiful Ada was the first **computer programmer**.

We show you how to bluff your way through modern computer programming in Chapters 5-7 and introduce you (gently) to the concept of a high-level computer programming language: it is very pleasing to note that PASCAL, BABBAGE and ADA are all now immortalised in the names given to three different high-level programming languages. We suggest, incidentally, that there is no finer way for the Computer Bluffer to show his or her firm commitment to the reality of the computer revolution than to add these illustrious names to those of any children he or she may currently be about to name: Kevin Babbage Bloggs, for example, Tracy Ada Buggins or Joe Pascal Doe, will we confidently predict, add a certain sound technology *cachet* to the lives of currently new infants when they later grow to maturity in the twenty-first century ...

... Each 1,024 consecutive bytes of STORE can be thought of as a conceptually complete 'block'. This block is called 'one thousand' bytes, or a 'Kilo Byte', abbreviated simply to 'KByte'. We thus have the

**Second Very Important Thing To Remember (SVITTR):**
**1K = 1,024**

This SVITTR neatly encapsulates all the concepts of data, store, process, address and register. It is simple, and powerful, bluff material and may easily be learned ...

## The Modularity Principle

Adopt a modular structure approach to analysing and designing the processes of your system. In other words, split your overall system down into the smallest self-contained units of processing as possible, each made up of sufficient elemental steps to define completely a logically complete task of the system. When you have done this, distinguish between inputs to processes which are parameters (usually small in number, with a limited range of values, each of which represents a whole class of data) and those which are the data themselves. Identify precisely all the inputs, outputs, logic processes (decision tables), and subprocesses within the module. Write down, in English, the elemental steps in the process—its complete 'recipe'.

This last Principle, the Modularity Principle, actually serves as the one which, if held to, ensures that the rest should be followed. For we have a special word for the 'recipe' of a process module. And it is, in fact, the Third, and Last, Very Important Thing To Remember (and undoubtedly the Most Important) in the Whole Wide World of Computer Bluff:

### TVITTR: ALGORITHM

An algorithm is the closely-defined set of steps describing precisely what our process module should do. In general parlance, it is the collection of exact procedural steps, together with the control statements, which specifically delineate a given task. It goes without saying, therefore, that the inputs and outputs required, together with any intermediate results, are thereby also precisely defined.

The simplest example of an algorithm in everyday life is perhaps the cookery recipe: all the inputs (ingredients) are defined, as are the processing steps (add ingredients, mix, cook) and the outputs (colour, aroma, number of helpings). Notice that control statements closely define any logical decisions to be taken. ('If the meringues at the top of the oven are tending to go brown, exchange them with

*. . . recipe for success*

those at the bottom, cooler part of the oven; if not, leave well alone').

   Another example of an algorithm would be the specification of the step-by-step procedure to be followed to *sort* the numbers 10,3,24,1,5,7 into ascending numerical order; then to find their sum; and then their arithmetic mean (average). Computer Bluffers with just a little arithmetical facility may find this exercise quite easy:

   inputs: 10,3,24,1,5,7
   outputs: 1,3,5,7,10,24;   (sorted into ascending
                            numerical order)
   $1+3+5+7+10+24 = 50$;   (summed)
   $50/6 = 8.333333$.      (average—a recurring
                            decimal).

It may not be so easy to write down a precise sequence of steps to specify exactly how to get from the inputs to the outputs (try it!). The point about designing, discovering or in some way hitting upon the algorithm to do this is that we are likely to find that we can *generalise* it to cater for any number of input numbers in any starting numerical order. Such a generalised algorithm is often referred to as a *routine or subroutine*.

# Irony as a phenomenon in natural science and human affairs

*R V Jones*

Ludwig Mond Lecture presented at the University of Manchester 29 February 1968. *Chemistry and Industry* 13 April 1968. Condensed.

Somewhere in his writings Sir Winston Churchill commented on the paradox of the Pyramids. The Egyptians, he pointed out, took monumental care to ensure that their Pharaohs should be entombed in the most substantial structures of the Ancient World, so that their mummies should remain inviolate: and yet this very action was in itself to attract the attention of posterity to the tombs and thus ensure that the remains of the Pharaohs would be disturbed. It demonstrates not only that 'the best laid schemes of mice and men gang aft agley' but also that this sometimes occurs because of something inherent in the schemes themselves. Such mishaps are, in fact, so frequent that it may be worth drawing some of the examples together to see whether there is a common pattern; and this is the opportunity that I propose to take in the Ludwig Mond Lecture. I have selected 'Irony' as the title because one of its meanings covers exactly what I wish to describe: 'A contradictory outcome of events as if in mockery of the promise and fitness of things'.

It would make a happy start in this context if I could point to some such irony in the life of Ludwig Mond. The nearest, perhaps, that I can get to it is his choice of nickel for the stopcocks in his chlorine plant at Winnington, because he had found that in the laboratory pure nickel resisted the attack of ammonium chloride vapour. In his large scale plant, however, he found that the nickel cocks became pitted and covered by a black crust of carbon. He soon found that these effects were due to a trace of carbon monoxide in the nitrogen with which he scoured his apparatus, and that the monoxide unexpectedly combined with nickel to give gaseous nickel carbonyl. Having thus given, in Kelvin's words, 'wings to a heavy metal' he used the fact, originally a bugbear, to develop a new process for making very pure nickel. In summary, having chosen nickel for its incorrodibility, he discovered that it was almost uniquely corrodible; and he went on to use his discovery in a process that was ultimately far more important than the process which he had originally been trying

36

to exploit. As my uncle used to teach me at shove-ha'penny, there are always two shots on the board—the one you aim for and the one you get.

Sometimes the shot you get is better than the one you expected. There is a story that P G Tait, Professor of Natural Philosophy at Edinburgh, and who was profoundly interested in golf, calculated how far a man could hit a golf ball, and that his son promptly took him out and showed him that a ball could be driven substantially further. P G Tait attributed the extra range to bottom spin. In this he was followed by J J Thomson, but the main cause may have been something different, for it turned out unexpectedly that old golf balls went further than new ones; and this was ultimately traced to the fact that old balls were dented and chipped. The air flow around a smooth ball at first follows the surface of the ball, but it ultimately breaks away, leaving great turbulent eddies in the wake. The energy in these eddies comes from the original forward kinetic energy of the ball, which thereby rapidly loses forward speed. With a rough ball, the irregularities on the surface cause turbulence to start in the air flow at an earlier stage than that at which it starts with a smooth ball, but the eddies although starting earlier stay smaller as they circulate around the ball—it is as though a little turbulence earlier in life is a way of ensuring that greater turbulence does not ensue later. The net result is that with a rough ball the over-all loss of turbulent energy is less than that with a smooth ball, and so the rough ball goes further. This is why golf balls are dimpled—an example of technological irony that can, incidentally, explain some of the mysteries of seam bowling at cricket.

These introductory examples will, I hope, indicate my theme. Irony is evidently a general phenomenon spanning many aspects of human affairs—administration, psychology, economics, education, security and war—as well as science and technology. We have had such ironies in the history of thought when different facets of the same truth have been glimpsed by different observers and when controversy has ensued because neither observer could see that the two facets were actually consistent. Eighteenth century physicists, for example, were divided regarding the nature of electricity; one school led by Franklin and Aepinus, argued that it consisted of a single charged fluid that moved through the interstices of a metal. Coulomb, in the other school, held that there were two fluids of opposite charge. There was intense, controversy, but one can now see that each side was in its way right. There are indeed two kinds of charge, but in a metal it is only the negative one that can move freely as a fluid.

Again, in the history of physics we had the wave-particle controversy—what could seem more different from a discrete particle than a diffuse wave? Wars have been fought for smaller differences than this, and yet the fact is that the structure of the natural world seems to be an harmonious synthesis of these apparently contradictory properties in one entity. It is therefore always worth standing back from any controversy which can be worthily supported from two sides to see whether the truth is really an unforeseen synthesis of the two viewpoints.

While the ironies of science are, perhaps, delicate and subtle, those of technology are often robust and tangible. They occur when an apparently good creative idea somehow produces the opposite effect to that intended by the inventor. Consider, for example, the plight of the railways in Lapland, where train operation was beset by the hazard of reindeer wandering across the track. Someone had the ingenuity to equip the trains with powerful loudspeakers emitting wolf calls, with the idea of scaring the reindeer away. In this it was successful, but the final state was no better than the first, since the track was then infested by wolves.

I was once the witness of what might be described as a colossal irony arising from one of my lectures, when I had shown that, owing to the incompressibility of water, one could literally shatter a filled beaker by firing a pistol bullet into it. The demonstration was taken so effectively to heart that its significance was transmitted to the Territorial Detachment of the Corps of Royal Engineers stationed in Aberdeen. This Detachment gladly sought tasks that would be helpful to the community and which would provide it with opportunities for practising its peculiar skills. One such exercise that it pursued from time to time was the felling of disused chimneys. The normal procedure in such instances was to remove some of the bricks from one side of the chimney near its base, and to replace the bricks by wooden props to help take the weight of the upper part of the chimney. When enough bricks had been removed, a fire was then lit in the base of the chimney, burning through the props, and thus causing the chimney to fall. When the next chimney had to be felled, however, it occurred to the engineers that a more effective way of demolition would be to stop up the hole at the base of the chimney and to fill it to a height of about six feet with water and then to simulate my pistol shot by firing a charge inside the water. The day chosen for the event was, inevitably, the Sabbath, since this was the usual Territorial operating day, and the news of the imminent spectacle did much to brighten the dullness of an Aberdeen Sabbath. The crowd assembled, the charge was fired, and the process entirely failed because it worked too perfectly. What happened was that every brick up to the height of the water was immediately blown outwards, leaving a chimney with a precisely trimmed base six feet up in the air. The whole chimney proceeded to fall intact, dropping neatly on to the level foundation below, and so remaining just as upright as it ever had been. Here was a truly 'contradictory outcome of events', the chimney now standing much more dangerously than before 'as if in mockery of the promise and fitness of things'.

Technical ironies occur with peculiar force in warfare. When, for example, our bomber losses began to mount in our 1943 night offensive against Germany, we equipped our aircraft with rear-ward looking radar sets to give warning of the approach of German nightfighters. If anything this probably increased our losses, because a radar set can be detected from other aircraft at a much greater range than that at which this aircraft can be detected by the radar set, and we thereby enabled German nightfighters to home on to our bombers from much greater distances.

Another kind of irony occurs when someone makes a show of efficiency by drawing up an elaborate specification, and when the very impressiveness, by its elaboration of detail, causes him to lose sight of the main object. The War Office in 1940, for example, had a specification for the ideal material with which to make crash helmets for motorcycle despatch riders. It included such items as the coefficient of elasticity, density, and so forth. One of my friends who was working on the design of the helmet found that, according to the specification, the ideal material from which to make it was sheet glass—the specification, despite or because of its length, had failed to mention that the material should not be brittle.

Sir Lindor Brown exposed a further vulnerability of specification-worship when he was Professor of Physiology at University College, London. He included in his annual request for equipment a sum of some hundreds of pounds for a digital frequency generator in the range 20 to 20,000 cycles/sec. The item survived inspection by all his colleagues at the College meeting in which the estimates of individual departments were ultimately cut down to fit the total sum made available by the finance committee, until he himself pointed out that what he had specified was, more simply, a grand piano.

Let us finally glance at some ironies in education. We have no time to look at the often remarked irony that selects professors for their originality and then so overloads them with administration that they have little time left for constructive thought in their own subjects. But it may well be worth mentioning another irony: examination standards can be set too high as well as too low. My friend Professor Yves Rocard blamed this phenomenon as part of the cause of the decline of French physics between 1900 and 1940; he said that examination standards became so high that instead of these resulting in a stream of better trained physicists, they discouraged many able students from even starting on university courses, and so the numbers of French physicists fell.

Nationally we face another irony. The present drive for widespread education comes basically from a belief that science and technology are vital factors in our national survival and advancement. Unfortunately, our post-war Governments failed to appreciate in time some of the vital details; they absorbed too many scientists, particularly physicists, in government establishments. There were therefore too few left to teach in schools and universities... The extra places at universities have therefore gone by default to fields other than science and technology. And so the money that was in effect largely won by the exertions of 1940 and intended for producing more scientists and technologists relative to the rest of the university output is ironically going on producing more students in subjects for which the national need is not so clear.

We have seen how general a phenomenon irony is. Fresh examples will occur to all; I have not even catalogued all the types. We have, for example, a national weakness that I have not so far mentioned—that of concentrating on the difficult and clever parts of a project and of forgetting to attend to the more prosaic periphery. This is one reason why we failed to consolidate our early lead in computers.

But in recognising these awkward facts, let us not be discouraged. It would indeed be an ironic result of this lecture if I were to inhibit administrators with fear of the consequences of irony. Rather I would invite them to take human nature more into account, and to visualise themselves more in the positions of those whose lives they profess to regulate; the unhappier ironies may then be avoided.

# Prestige quantified

In their examination of scientific output and recognition—a study in the operation of the reward system in science—Stephen Cole and Jonathan R Cole asked 1300 university professors to rank 98 prizes and honors. They found that only 42 of these were known to as many as 20% of physicists and only 22 awards were known to as many as half of the physicists. (One might wish that names of one or two fictitious awards had been included in the questionnaire.)

The table of results starts as follows:

| Name of award | Mean prestige score | % Ranking the award (visibility score) |
|---|---|---|
| 1) Nobel Prize | 4.98 | 100 |
| 2) Enrico Fermi Award | 4.31 | 92 |
| 3) Member National Academy of Sciences | 4.22 | 95 |
| 4) Royal Astronomical Society Gold Medal | 4.20 | 30 |
| 5) Albert Einstein Gold Medal | 4.19 | 47 |
| 6) Member of Academie Francaise | 4.12 | 56 |
| 7) Fritz London Award | 4.03 | 65 |
| 8) National Medal of Science | 4.02 | 38 |
| 9) Membership, Royal Society, London | 4.01 | 86 |
| 10) Lorentz Medal | 3.98 | 36 |
| 11) Max Planck Medal | 3.97 | 47 |
| 12) Presidential Medal of Freedom | 3.86 | 43 |

# Continents adrift: new orthodoxy or persuasive joker?

[ Reproduced by permission (RBP) from R S Dietz and J C Holden 1973 *Implications of Continental Drift to the Earth Sciences* vol 2 ed D H Tarling and S K Runcorn, NATO Advanced Study Institute, April 1972, University of Newcastle-upon-Tyne (London: Academic Press). Copyright © 1973 by Academic Press Inc. ]

The modern rebirth of interest in continental drift commenced, we believe, in the mid-1950s when it was shown by a certain British geophysicist (who will remain nameless) from studies of rock magnetism that the polar-wander curves of North America and Europe, congruent in earlier times, diverged by about 30° in the Cretaceous and Cenozoic. This was widely, or perhaps wildly, hailed as proving the opening of the North Atlantic by continental drift.

This enthusiasm proved premature, because geologists rather carelessly handle their rocks, as Sir Harold Jeffries (1970) recently pointed out in his book, *The Earth*. He writes:

> 'When I last did a magnetic experiment (about 1909), we were warned against careless handling of permanent magnets, and the magnetism was liable to change without much carelessness. In studying the magnetism of rock the specimen has to be broken off with a geological hammer . . . It is supposed that, in the process, its magnetism does not change to any important extent, and, though I have often asked how this comes to be the case; I have never received any answer.'

For this, and other reasons, Jeffries dismisses Wegener's drift as being 'quantitatively insufficient and qualitatively inapplicable. It is an explanation which explains nothing which we wish to explain.' It would seem, therefore, that rocks collected for paleomagnetic measurements by hammers (and how else does a geologist collect his rocks) would have their magnetic memory scrambled. This being obviously so, rock magnetic studies must be regarded as worthless. The first law of geology, we suppose, is: The rocks remember, while liquids and gases forget. But it can hardly apply to rocks which have been hit on the head.

On the authority of Beloussov (1962) we have it that, 'the hypotheses suggesting horizontal drift of the continents, among them the hypothesis of Wegener, which was once famous, must be regarded as fantastic and having nothing to do with science . . . It is a source of profound amazement that such a hypothesis—based as it is on an overtly formalistic approach to the major problems and on a total and consistent disregard of the basic geotectonic data and, as already stated, explaining nothing of what must be explained in the first place—was not only seriously discussed in scientific literature but achieved considerable success and attracted some of the leading authorities into the ranks of its adherents. These men were apparently hypnotized by the boldness of Wegener's ideas and by his brilliant style of writing.'

This was a traumatic revelation to the zealots of continental drift and they fell into disarray. Like all true scientists, the drifters would 'rather be right than President,' but obviously they were neither.

## The Animals Remember

All was not lost, however, for new evidence came in from an entirely unexpected quarter—the animal kingdom. Animals have some remarkably developed instincts, which sometimes recapitulate their evolutionary history. Witness the so-called loud bats with their FM sonar chirps by which they can search out and classify a moving target—usually moths, their favourite meal. But it remained for a bird, by its remarkable migratory path, to first demonstrate that the New World really has drifted away from the Old World.

This doughty bird, the sooty hoodwink, *Puffinus oceanicus*, winters in the Atlantic sector of Antarctica; then each spring it heads north, determined to nest in far away Spitzbergen. As if flying to this remote island is not a sufficient demonstration of fortitude, this bird chooses a zigzag path. First it touches down in Southwest Africa where, because of its confused and dazed habit of stumbling about (apparently searching for fresh bearings), it is locally termed the random walkabout. Then this bird executes further zigs and zags across the ocean as it threads its way north. On April 1st, the frayed remnants of the flock touch down on the British Isles at Lands End. (Remarkable as it is, their navigation sometimes goes awry. An errant flock was seen in 1967 far off course in the spaghetti fields of the Po valley.) Toward the end of April the sooty hoodwink finally reaches its destination, Spitzbergen.

FIGURE 1
... bat sonar

44

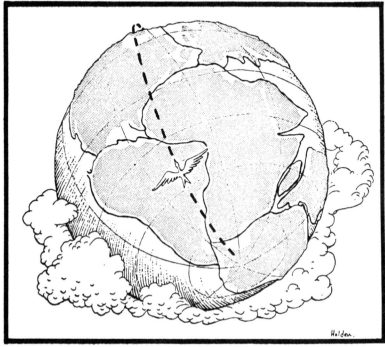

FIGURE 2 . . . a doughty bird

45

This curious migration path finds ready explanation once we recognize that continental drift has occurred. If we close the Atlantic Ocean, we find that the zigzag path becomes a straight line, a great circle route, or the shortest distance between Antarctica and Spitzbergen.

Another animal, albeit extinct, fills in still another facet of the continental drift puzzle. We refer here to *Glossopstompodon loathifoliata*, whose bleached remains are found in the Permian red beds of the Sahara. Critics of drift have argued that there never could have been a universal continent of Pangaea in the late Paleozoic, because, if this were true, the Glossopoteris flora on Gondwana would certainly have invaded the northern continents of Laurasia. These critics, however, failed to reckon with the tempestuous temper and unremitting phobia of *Glossopstompodon* for glossopterids. By setting up a rampaging patrol along the equator and trampling any young glossopterids sprouts, this reptile established a successful barrier against the northward migration of this flora. (We are reminded here in passing of the acceptance speech by a young English geophysicist when receiving a medal for his contribution to plate tectonics. If I recall correctly, he said with characteristic British modesty, 'If I have seen farther than others, it is because I have stomped on heads of giants.')

A somewhat similar explanation applies to the curious swimming behavior of the deep sea squid, *Architeuthis solenoides*. Nearly all squid swim backwards through life, apparently preferring not to look where they are going, but to see where they have been. By swimming forward, *A. solenoides* is exceptional, but apparently he has not always swum in this manner. Experiments show that he may be programmed to swim either forward or backward within an aquarium surrounded by a coil simply by

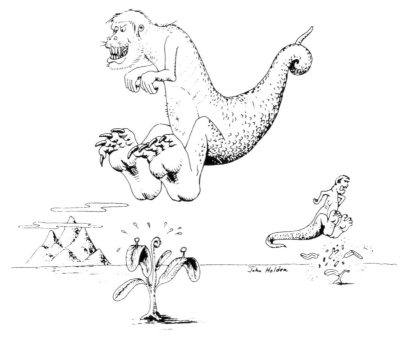

FIGURE 3
. . . *Glossostompodon loathifoliata*

46

FIGURE 4

... *Architeuthis solenoides*

reversing the magnetic field. It would seem, then, that *A. solenoides* became a forward swimmer only 700,000 years ago when the earth's field switched from reversed to normal at the Matsuyama–Brunhes boundary.

We have learned of late from the paleomagicians that, although 'east is east and west is west, and never the twain shall meet,' this adage does not apply to north and south. Every so often, and quicker than you can say Willem Jean Marie van Water-schoot van der Gracht, north may become south and, we hope, vice versa. We have learned this from the 'fossil compasses' frozen in basalts and other rocks. Actually, animals which are senstive to the magnetic field of force have known this all along. It has been known for many years that, upon molting, crabs place a grain of sand in their inner ear, which then becomes a sensor for geotropism or, more simply, balance. Some crabs carelessly choose a grain of magnetite—a ferromagnetic mineral. Then, if a strong magnet is above their aquarium, they will forever crawl along the roof of their home. This much is evident by direct experimentation. But why do crabs crawl sideways? This mystery is solved when we recall the frequent flips in the earth dipole field. The crab becomes confused as to whether he should walk forwards or backwards, so, adapting to compromise, he instead walks sideways.

Yet another animal may be cited as proving continents drift. This is the common European eel, *Anguila.* After a few years in the streams of Europe, the eel heads for the Sargasso Sea on the far side of the mid-ocean spreading rift to spawn and then die. Then the newborn eel, the *Leptocephalus* stage, swims back to Europe—this instinct inherited from its parents. But by this time Europe is not where it was supposed to be,

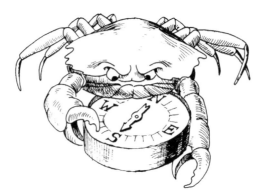

FIGURE 5
. . . and never the twain shall meet ( ?)

FIGURE 6
. . . one more last step

as this continent has drifted several centimeters eastward. No one knows for sure just what crosses the mind of a young eel in the face of this predicament. All we know for sure is that all eels go through life with a quizzical look. A chinese proverb runs that, in a journey of a thousand miles, the last step is no more important than any other. But suppose that every year someone adds *one more* last step. For eels, it would seem that one more step beyond the last step is needed for survival of the species.

A final animal which adds credence to drift is none other than the famous Loch Ness monster. Photos and descriptions of this elusive monster show its long neck and small, reptilian head, so that it is a swimming 'dinosaur' or pleisiosaur, a living fossil from the Jurassic. It also has been noted by boats plying the loch that Nessie yields to the right, thus obeying Napoleonic (or North American) rather than Caesarian (or British) rule-of-the-road. Clearly, then, Nessie is a beast of the New World now stranded in the Old World by the break away of Europe from North America in the Cretaceous. There seems now to be a new urgency for somehow corralling Nessie and, while treating her with the tender loving care that befits an endangered species, making such measurements as would permit identification as to species—for example, counting her teeth.

FIGURE 7
. . over the loch from Urquhart castle

## Historical Beginnings

Before proceeding further, we would like to pay homage to the originators of continental drift. There seems little doubt that the concept is originally ascribable to Frank Bacon (Blackett, *et al.* 1965), who, in his *Novum Organum* of 1620, wrote: 'The very configuration of the world itself in its greater part presents Conformable Instances which are not to be neglected. Take, for example, Africa and the region of Peru with the continent stretching to the Straits of Magellan, in each of which tracts there are similar isthmuses and similar promontories, which can hardly be by accident.' Bacon undoubtedly thought that, given this hint, the reader would have sufficient intelligence to see that the South Atlantic Ocean can be closed in a continental drift reconstruction. This, of course, is simply a matter of flipping South America upside down, north for south, and then sliding this continent eastward such that the bulge of Peru fits beneath the bulge of Africa. (We should mention in passing that there is no truth to the rumor that Bacon wrote the Shakespearean plays. For that matter, Bill Shakespeare didn't either. They were written by another man of the same name—and that should be an object lesson in general semantics.)

Let us also set the record straight as to the first symposium on continental drift, because A. Meyerhoff (1972) accorded that niche in history to the American Association of Petroleum Geologists 1926 symposium on continental drift organized by W. J. M. van Waterschoot van der Gracht. This is incorrect. The distinction clearly belongs to Samuel Pepys, FRS.

Pepys' diary for 23 May 1661 reads:

'To the Rhenish wine-house in Crooked Lane, and there Mr. Jonas Moore, to us, and there he did by discourse make us fully believe that *England and France were once the same continent* (italics added), by very good arguments, and spoke very many things not so much to prove the Scripture false, as that the time therein is not well computed nor understood. In my black silk suit (the first day I have put it on this year) to my Lord Mayor's by coach, with a great deal of honourable company, and great entertainment. At table I had very good discourse with Mr. Ashmore, wherein he did assure me that frogs and many insects do often fall from the sky, ready formed.'

49

From the above, it is clear that Pepys accepted the view that England had drifted away from Europe. A symposium in the purest sense of that word, and, as its Greek roots reveal, is a wine-drinking party. To Samuel Pepys, Cheers!, or Drink Hail!— for convening the first symposium on continental drift.

## Humpty Dumpty had a Great Fall

Piecing together all of the continents has been described as a jigsaw puzzle. This is hardly correct, because solving a jigsaw puzzle is child's play. In contrast, piecing the continents together would seem to be a Humpty Dumpty problem, for, as you will recall, all the king's horses and all the king's men could not put Humpty Dumpty together again. Similarly, all the world's eggheads have been unable to reconstruct Pangaea.

The congruency of the margins of Africa and South America has always been the inspiration for drift. Alfred Wegener (1922) opened his classical book on drift with: 'He who examines the opposite coasts of the South Atlantic Ocean must be somewhat struck by the similarity of the shapes of the coast line of Brazil and Africa . . . . This phenomenon was the starting point of a new conception . . . called displacement of continents.' Wegener apparently did not realize that many decades ago A. Snider (1859) had already quantitized this fit with nice precision. This is presented in his remarkable book—'The Creation and its Mysteries Revealed: A Work which Clearly Explains Everything Including the Origin of the Primitive Inhabitants of America, etc., etc.' Snider illustrated by lithograph the fit of the New World against the Old World (see cut). It will be noted that, in his closing of the Atlantic Ocean, the match is

FIGURE 8
. . . Snider's fit of 1858 BC

perfect with neither overlaps nor underlaps. This fit is far superior to those attained by recent workers—for example, the Bullard *et al.* fit (1965). And Snider's fit was obtained in 1859 BC—yes, BC, Before Computers.

Subsequently, Carey (1958) has redone the Snider fit with cartographic precision, juxtaposing the 1000 fm isobaths on a common stereographic projection. This shows that the knee of South America fits snugly into the groin of Africa, that broad swales match broad re-entrant, bumps fit into bays, and even bumps on the bumps fit into bights in the bays. Of Carey's fit, Chester Longwell (1958) remarked, 'If the fit between South America and Africa (is not a gigantic rift), surely it is a device of Satan for our frustration.' Unlike the zealous drifters, Longwell at least has given us a choice. In view of this dilemma, we suppose we must resort to higher authority. In this respect, we can hardly do better than the eminent Sir Harold Jeffries (1970), who writes: 'On a moment's examination, the alleged fit of South America into the angle of Africa is seen to be really a misfit by about 12°.' It would seem that this supposed fit is no more than a persuasive joker.

Perhaps no one has shown better than Meyerhoff & Meyerhoff (1972) the folly of attempting to shoehorn continents together. In a diagram showing a family of squiggles said to represent the outline of Japan he has shown that Japan may be fitted almost anywhere in the world. Another masterwork of meyerhoffiana reveals that the eastern margin of North America, when turned upside down, fits nicely against the eastern coast of Australia. There has been much argumentation over the years about the position of Madagascar when the Indian Ocean is closed—whether this microcontinent fits against Tanzania or against Mozambique. In point of fact, both positions are morphologically poor. A proper fit is achieved only when Madagascar is leapfrogged over Asia and inserted into the Caspian Sea.

## Wheels Within Wheels Within Wheels

It has been widely proposed that the continents can be piggybacked about by convection of the mantle. With convection one can do almost anything as the entire process is wonderfully amenable to mathematic manipulation. And there are many modes of convections—toroids, plumes with thunder-heads, helixes, etc.—all of which can be readily explained by arm waving which conjures up explicit models. Furthermore, simple but ingenious experiments can be performed at which the British scientists excel. The result has been a new third school of experimentalists (the earlier schools being the 'baling wire and sealing wax school' and the 'negative experiment school') which may be termed the 'kitchen experiment school'. But in all fairness it should be pointed out that this last-named school was anticipated in America by those early workers who successfully modeled lunar craters by dropping marbles into porridge. And we must mention here the important contribution by an early selenologist on the American frontier who discovered nearly a century ago that, regardless of the obliquity of the angle at which he shot a buffalo, the hole, like a lunar crater, was always round.

In defense of convection, a dimensionless formula is presented (see cut) for oboe and flute. Ideas that are too bizarre to record in writing may yet be sung with perfect propriety. The equation has no particular relevance to the present discussion, but it does add a touch of elegance. In this particular equation, it will be noted that all of

FIGURE 9
... a proper fit

FIGURE 10
... wheels within wheels within wheels

the terms are either negligible or trivial; hence the entire equation is inconsequential. If we are ever to deal successfully mathematically with convection in the mantle, we must first establish some simplifying boundary conditions. For example, the rings around Saturn remained beyond understanding until a certain German mathematician cracked the problem by assuming just two simplifying conditions, the first being that the rings are square and the second being that the rings are at an infinite distance from Saturn.

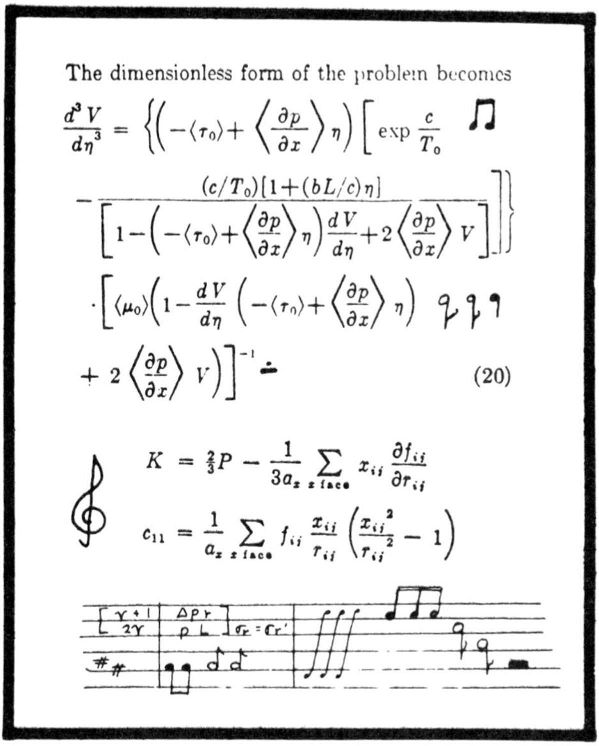

The dimensionless form of the problem becomes

$$\frac{d^3 V}{d\eta^3} = \left\{ \left( -\langle \tau_0 \rangle + \left\langle \frac{\partial p}{\partial x} \right\rangle \eta \right) \left[ \exp \frac{c}{T_0} \right. \right.$$

$$\left. - \frac{(c/T_0)[1+(bL/c)\eta]}{\left[ 1 - \left( -\langle \tau_0 \rangle + \left\langle \frac{\partial p}{\partial x} \right\rangle \eta \right) \frac{dV}{d\eta} + 2 \left\langle \frac{\partial p}{\partial x} \right\rangle V \right]} \right]$$

$$\cdot \left[ \langle \mu_0 \rangle \left( 1 - \frac{dV}{d\eta} \left( -\langle \tau_0 \rangle + \left\langle \frac{\partial p}{\partial x} \right\rangle \eta \right) \right.$$

$$\left. \left. + 2 \left\langle \frac{\partial p}{\partial x} \right\rangle V \right) \right]^{-1} \doteq \qquad (20)$$

$$K = \tfrac{2}{3} P - \frac{1}{3 a_x} \sum_{x \text{ face}} x_{ij} \frac{\partial f_{ij}}{\partial r_{ij}}$$

$$c_{11} = \frac{1}{a_x} \sum_{x \text{ face}} f_{ij} \frac{x_{ij}}{r_{ij}} \left( \frac{x_{ij}^2}{r_{ij}^2} - 1 \right)$$

FIGURE 11
. . . dimensionless formula for oboe and flute

## Continents on the Move

We are told that the earth's crust is a mosaic of six large, 100 km thick plates in relative motion and that the continents are embedded in these plates as passive passengers. The plates never impinge, for they subduct instead, but the continents do, for they are composed of sial which is not geo-degradable. And it is these continental collisions which cause orogenies—linear mountain belts thrown up along the margins of continents. The Miocene collision of the Indian subcontinent with the soft underbelly of Asia is the type example. The three-mile-high Himalayan rampart overlooking the Gangetic plain resulted and India received a bashed-up nose in the process. We may liken this to the English bulldog, renowned for his remarkable tenacity, and some say, but I am sure it cannot be true, stupidity. Anyway, the comparison holds, because the abrupt profile of the English bulldog comes from chasing after parked cars.

A tenet of plate tectonics is that the ocean basins are ephemeral features—they are either opening or closing, but the continents exist forever—suturing up, unsuturing and generally being shunted about. The Pacific plate is moving fast enough to entirely circumnavigate the globe during the Phanerozoic alone. Africa collided with North America in the Devonian 400 m.y. ago, triggering the abrupt Acadian orogeny. This collision must have stored elastic stresses, for 180 m.y. ago Africa took off again in the opposite direction, breaking apart generally along the earlier suture. In view of this historical scenario, we can expect Africa's return at any time, as the Atlantic Ocean cannot grow forever larger and larger. Surely a new plate boundary will form along the American shore and, first of all, a massive slab of the upper mantle will be obducted or thrust across the eastern seaboard, sweeping away the cities like some giant glacier. The implications are catastrophic; time is of the essence and we need now to establish a Continental Drift Early Warning System.

FIGURE 12
. . . Continents adrift

## A New Twist

It is now well known that the Sea of Tethys, a vast arm off the former universal ocean of Panthalassa and lying between Gondwana and Laurasia, was the site of a subduction zone or trench along which there was also considerable transform slippage which has been termed the *Tethyan twist* (van Hilton, 1964). This slippage apparently was sinistral in the early opening phases of the North Atlantic Ocean, but became dextral during the Cenozoic, carrying Spain back from the open Atlantic to its present point of contact against Africa.

The Tethyan twist was one of the major scenarios of drift, but perhaps an even more important one, undiscovered until now, was the sinistral *Equatorial twist*. The proof that this twist has occurred lies in the remarkable symmetries achieved. The transformation may be performed using a suitable globe (e.g., the National Geographic 16″ Physiographic Globe) by slipping the northern hemisphere of the earth along the equator dextrally for precisely one quadrant of earth (90°). The result is an

54

entirely new distribution pattern of land and oceans, but one which is entirely realistic. Indonesia is found to be exactly on register with the northern half of Africa, producing the new continent of Afrodesia. The mid-ocean rift of the North Atlantic lines up precisely with that of the southern Indian Ocean, producing an Indo-Atlantic rift. The southern half of Africa falls on register with the northern half of South America, producing an Amerafrica. Farther west, the North Pacific ridge can be connected to the South Atlantic ridge by imposing only a short segment of ridge-ridge transform fault, forming the Paclantic ridge. Proceeding even farther west, one finds the southern portion of South America standing alone in the central Pacific. The skeptic may pounce upon this as an oddity which vitiates entirely our Equatorial twist. But the reverse is true, for this is the lost continent of Mu whose former existence was adduced many years ago by James Churchward. This gentleman found the history of Mu written on secret tablets of stone amongst the archeological ruins of Mexico. He erred in just one respect: Mu did not sink beneath the waves of the Pacific, but instead it suddenly was wrenched eastward to become the southern portion of South America. It seems that some eons ago an errant asteroid, a cosmic cannonball, zeroed in on the earth, and . . . Bang!

The equatorial twist provides no answers, but it poses many questions.

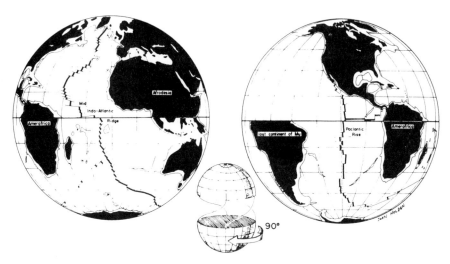

FIGURE 13
. . . the Equatorial Twist

## Expansive Thoughts

There is a school of natural philosophy, prominent in the topsy-turvy world of Down Under, that our planet was once much smaller than it is now—and that it expanded to its present size only in the past 200 m.y. A common version of this concept is that the continents were all once joined and even the Pacific was closed, so that the world was without ocean basins. Continents, to the 1000 fm isobath, cover 40% of the earth, so that this earth would have only two-fifths the area of the modern world ($510 \times 10^6$ km²). The 'drifting' of continents could then be explained in a manner somewhat similar to the dispersion of points on the surface of an expanding

FIGURE 14
. . . small world of the Triassic

balloon—or like the stars in our expanding universe. However, only the oceans would grow while the continents maintained their original dimensions. The continents then would separate by displacement rather than drift.

Animals once again provide a key and appear to demonstrate that the earth may have been smaller in the Triassic than now. Assuming no change in Big G and accomplishing this miniaturization by phase change, the world of the early Triassic would have had a density of 22 and animals roaming the surface would have weighed $2\frac{1}{2}$ times as much as now. But this is not too unreasonable. Witness the squatty robust morphology of the fossil tetrapod, *Lystrosaurus*, with his belly scraping the ground. Does this stance not suggest that this sheep-sized reptile weighed not 100 pounds (scaled to the modern-sized earth), but actually 250 pounds in the *Triassic*? And why do you suppose all of those deep tracks and trails are found impressed on Triassic red beds. Their greater weight on a smaller earth would seem to provide the answer.

Many other things would happen on a 40% earth. Lazy rivers would become torrents. The nearly tripling of the Coriolis force would cause animals unconsciously to roam in circles of ever increasing concentricity, finally chasing their own tails. The high gravity field probably explains why plants of the Triassic such as the horsetail (Equisetum) were stiffened with stems high in silica. Conservation of the moment of inertia would increase the rate of the earth's spin, so that the Triassic day would have been a mere ten hours long. This may expain why the modern descendants of the Triassic reptiles remain today as the sleepiest of all God's creatures.

Admittedly, for all its advantages there are some aspects of the expanding earth concept which need a bit of tidying up. For example, what to do with all that water if there were no ocean basins to contain that stuff—$1300 \times 10^6$ km². One solution would be to assume that it was all locked up as fresh water in enormous ice caps on the continents and, of couse, there was a great Permo-Carboniferous glaciation. But then what about all that residual salt. . . .

56

'I have discovered the length of the sea serpent, the price of the priceless and the square of the hippopotamus . . . and how many birds you can catch with the salt in the ocean—187,796,132 if it would interest you to know,' said the Royal Mathematician. 'There aren't that many birds,' said the King. 'I didn't say there were,' said the Royal Mathematician.

(Thurber/*Many Moons*)

Considering sodium chloride alone, the only type useful for taming birds, at 85% of total ocean salts we derive a figure of $43 \times 10^{25}$ tons. Rounding the number of birds to be tamed to $188 \times 10^6$ provides us with $230 \times 10^6$ tons per bird. With due regard for the Royal Mathematician's wisdom, this does seem a bit much to pour on the tail of each bird.

Carey (1970) would configure the world into eight primitive (Paleozoic) polygons— one for each continent plus an Eo-Pacific polygon, all of which would have moved away from each other over the past 200 m.y., causing a 75% increase ($220 \times 10^6$ km²) in the earth's surface area. This is a story of separating polygons, beginning with a roughly 60% earth, motivated by six convecting mantle toroids plus a jet stream. The old moribund passivity is replaced by a kaleidoscopic crust with gross churnings of the earth's interior. This may seem bizarre, but, as Carey points out, so was Wegener's drift a few years ago when 'the American bandwagon chanted *Ein Marchen*, a pipe dream, and beautiful fantasy.'

Carey adheres to the Egyed principle that the continents have dried off as the ocean was slowly withdrawn into the growing ocean basins. The ultimate cause, according to Egyed, was an expanding atom. The collapsed atoms of the inner core, he supposed, eventually evolved into outer-core substance with constrained electron shells and then into normal mantle rock. In other words, the core evolved into the mantle rather than the core being differentiated from the mantle as is usually supposed.

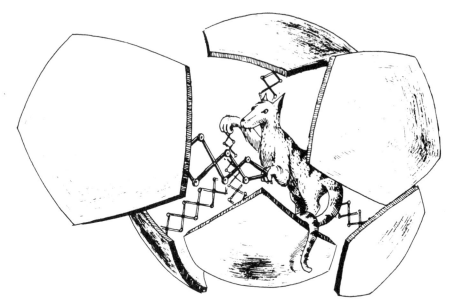

FIGURE 15
. . . Hungarian atom and the Tasmanian tesselated earth

57

Possibly Egyed of Budapest has provided ultimate key with which to unsuture Carey's zipped-up, tesselated earth of the Paleozoic. A Hungarian expanding atom would force apart the Tasmanian polygonal shells, causing the earth to grow and grow . . . .

In short, an expanding earth remains an exhilarating concept worthy of philosophic contemplation. It seems rather unkind that some unimaginative persons call it 'balloonus balonus.'

## Drifters are Alive, . . . and Well

We drifters are now being pushed around, even as we push the continents around. There is a need to be more assertive—and to make some more flat-out bold statements. We need a return to the faith of Galileo Galilei, who, when forced under duress to recant his theory that the Earth rotates on its axis and revolves around the Sun, muttered aloud as he left the Inquisition, 'Eppur si muove' ('but still it moves'). Earlier, in his Sidereus Nuntius of 1610, Galileo wrote, 'The Moon is not perfectly smooth . . . but, on the contrary, it is full of inequities, uneven, full of hollows and protruberances just like the Earth itself.' The philosophers scoffed at this bizarre notion, for it was at variance with the writings of Aristotle. And even if the Moon did have irregularities, it must still be covered with a thick, smooth crystalline layer. Otherwise, how would the Moon so clearly reflect the map of the Earth on its face, with the dark region being our oceans and the light regions being our continents? Galileo retorted with an answer precisely suited to the merit of this argument. 'Let them be careful, for, if they provoke me too far, I will erect, on their crystalline shell, invisible crystalline mountains ten times as high as any I have yet described.'

So, let those fulminous forecasters of fixity take warning. If we devious disciples of drift are not permitted to move our continents a paltry few centimeters each year, then we will drift them several meters each day!

## The Geotectonics Creed

In closing, it would seem that continental drift, as derived from sea floor spreading, transform faulting, and plate tectonics, is the new orthodoxy. As statement of faith, the oath of office for the modern global tectonicist seems in order as is provided by the new Geotectonics Creed (Scharnberger & Kern, 1972), with apologies to the Council of Nicea.

'I believe in Plate Tectonics Almighty, Unifier of the Earth Sciences, and explanation of all things geological and geophysical; and in our Xavier LaPichon, revealer of relative motion, deduced from spreading rates about all ridges; Hypothesis of Hypothesis, Theory of Theory, Very Fact of Very Fact; deduced not assumed; Continents being of one unit with the Oceans, from which all plates spread; Which, when they encounter another plate and are subducted, go down in Benioff Zones, and are resorbed into the Aesthenosphere, and are made Mantle; and cause earthquakes foci also under Island Arcs; They soften and can flow; and at the Ridges magma rises again according to Vine and Matthews; and ascends into the Crust, and maketh symmetrical magnetic anomalies; and the sea floor shall spread again, with continents, to make both mountains and faults, Whose evolution shall have no end.

58

And I believe in Continental Drift, the Controller of the evolution of Life, Which proceedeth from Plate Tectonics and Sea-Floor Spreading; Which with Plate Tectonics and Sea-Floor Spreading together is worshipped and glorified; Which was spake of by Wegener; And I believe in one Seismic and Volcanistic pattern; I acknowledge one Cause for the deformation of rocks; And I patiently look for the eruption of new Ridges and the subduction of the Plates to come. Amen.'

## References

Beloussov, V. V., 1962. Basic Problems in Geotectonics. McGraw-Hill, London, 465 pp.

Blackett, P. M. S., Bullard, E. and Runcorn, S. K., 1965. A Symposium on Continental Drift. The Royal Society, London, 323 pp.

Bullard, E. C., Everett, J. and Smith, A. G., 1965. The fit of the continents around the Atlantic. *In:* Symposium on Continental Drift, *Phil. Soc.* 1066, *Roy. Soc.*, **258A**, 41–51.

Carey, S. W., 1958. Tectonic approach to continental drift. *In:* Continental Drift, A Symposium, pp. 177–358. Univ. Tasmania Press.

Carey, S. W., 1970. Australian, New Guinea and Melanesia in the current revolution in concept of the evolution of the earth, *Search*, **1**, No. 5, 178–89.

Jeffreys, H., 1970. The Earth: 5th edition. Cambridge Univ. Press, 525 pp.

Longwell, C. R., 1958. My estimate of the continental drift concept. *In:* Continental Drift, A Symposium, pp. 1–12. Univ. Tasmania Press.

Meyerhoff, A., 1972. Continental drift, *Geotimes*, **17**, No. 4, 34–6 (rev. of book by D. and M. Tarling).

Meyerhoff, A. A. and Meyerhoff, H. A., 1972. The new global tectonics: major inconsistencies, *Amer. Assoc. Petrol. Geol.*, **56**, 269–336.

Scharnberger, R. and Kern, E., 1972. Geotectonics creed, *Geotimes*, **17**, No. 1, 9–10.

Snider, A., 1859. La Creation et ses Mysteres devoiles. A. Franck and E. Dentu, Paris, 487 pp.

van Hilton, D., 1964. Evaluation of some geotectonic hypotheses by palaeomagnetism, *Tectonophysics*, **1**, 3–71.

Wegener, A., 1924. Origin of Continents and Oceans (English trans. 3rd edition by J. Skerl), 212 pp. Dutton and Co.

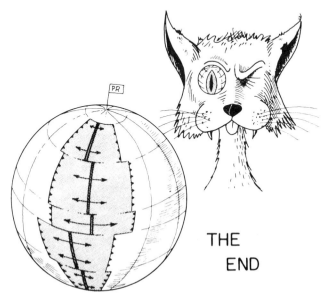

THE
END

. . . hypotheses, like cats, have nine lives

# From the devil's DP dictionary

Stan Kelly-Bootle
*The Devil's DP Dictionary* 1981
(New York: McGraw-Hill).
Reproduced with permission of McGraw-Hill Inc.

**algorithm** *n.* [Origin: ALGORISM with a pronounced LISP.] A rare species endangered by the industry's cavalier pursuit and gauche attempts at domestication.

♦ The current plight of the unspotted algorithm, *Algorithmus accuratus*, can be traced back to overculling in the 1960s. It will be recalled that the previous decade had witnessed an uncontrolled population growth, indeed a plague of the creatures in diverse academic terrains. Their pernicious invasion of the commercial environment in the late 1950s prompted IBM to offer the controversial $4.98 bounty per pelt. Hordes of greedy and unskilled people from all walks of life deserted their jobs and families, sold their possessions, and flocked to dubious, fly-by-night programming schools. Over-armed with high-level weapons, these roaming bands of bounty seekers hunted down and massacred the poor algorithm around the clock. The inevitable reaction occurred, but almost too late, in the form of a conservationist "Save the Algorithm" lobby, replete with badges, bumper stickers, and fund-raising algorithms. Public opinion was aroused, in particular, by a catchy campaign song:

> *Algorithm, algorithm, algorithm,*
> *Who could ask for anything more?*

**bit** *n. & adj.* [Origin: either Old English *bita* "something small or unimportant," or engineering *bit* "a boring tool."] **1** *n.* The quantum of misinformation. **2** *n.* One-half of the fee needed to carry out a threat, as: "For two bits I'd ram this board down your stupid throat." **3** *n.* A BINARY digit; a boringly dichotomic entity which precludes rational discussion. "Avoid situations which offer only two courses of action."—S. Murphy. **4** *adj.* (Of a programmer) inadequate; versed only in FORTRAN or RPG. **5** *adj.* (Of a map) many-1 and many-0. The 1s in a bit map indicate to the system those sectors of mass memory which are immune from further corruption.

60

**CAD** *n.* [Acronym for Computer-Aided Delay or, *archaic*, Computer-Aided Design.] The automation of the traditionally *manual* delays between the various stages of product development: research and development, drawing office, prototyping, testing, preproduction planning, etc. The improved delays invariably lead to better products.

**computer science** *n.* [Origin: possibly Prof. P. B. Fellgett's rhetorical question, "Is computer science?"] **1** A study akin to numerology and astrology, but lacking the precision of the former and the success of the latter. **2** The protracted value analysis of algorithms. **3** The costly enumeration of the obvious. **4** The boring art of coping with a large number of trivialities. **5** Tautology harnessed in the service of Man at the speed of light. **6** The Post-Turing decline in formal systems theory. "Science is to computer science as hydrodynamics is to plumbing."—Prof. M. Thümp.

**consultant** *n.* [From *con* "to defraud, dupe, swindle," or, possibly, French *con* (vulgar) "a person of little merit" +*sult* elliptical form of "insult."] A tipster disguised as an oracle, *especially* one who has learned to decamp at high speed in spite of the large briefcase and heavy wallet.

◆ The earliest literary reference appears to be the ninth-century Arabic tale *Ali Baba and the Forty Consultants.*

**cursor** *n.* [Possibly Old Irish *cursagim* "to blame" or English *cursory* "rapid, superficial."] A faintly flickering symbol on a CRT screen, used to test the eyesight and reflexes of the operator, and indicating where the next keyed character will be rejected.

◆ In parts of England frustrated terminal minders often refer to the *blinking* cursor. A cursor in the top left-hand position of an otherwise blank screen serves to indicate that the system (with the exception of the cursor-generation module) is inoperative. In well-designed systems the cursor flicker rate is set to match the operator's alpha brain rhythm to provide an inescapably hypnotic point of interest until normal service is resumed.

**FORTRAN** *n.* [Acronym for FORmula TRANslating system.] One of the earliest languages of any real height, level-wise, developed out of Speedcoding by Backus and Ziller for the IBM/704 in the mid-1950s in order to boost the sales of 80-column cards to engineers.

♦ In spite of regular improvements (including a recent option called STRUCTURE), it remains popular among engineers but despised elsewhere. Many rivals, with the benefit of hindsight, have crossed swords with the old workhorse! Yet FORTRAN gallops on, warts and all, more transportable than syphilis, fired by a bottomless pit of working subprograms. Lacking the compact power of APL, the intellectually satisfying elegance of ALGOL 68, the didactic incision of Pascal, and the spurned universality of PL/1, FORTRAN survives, nay, flourishes, thanks to a superior investmental inertia.

**random** *adj.* 1 (Of a number generator) predictable. 2 (Of an access method) unpredictable. 3 (Of a number) plucked from the drum, Tombola, by the flaky-fingered Tyche. 4 [From JARGON FILE.] (Of people, programs, systems, features) assorted, undistinguished, incoherent, inelegant, frivolous, fickle.

♦ *A mathematician in Reno,*
*Overcome by the heat and the vino,*
*Became quite unroulli*
*Expounding Bernoulli,*
*And was killed by the crowd playing Keno.*

The neo-Gideons are now placing copies of R. von Mises' *Wahrscheinlichkeitsrechnung* (Leipzig und Wien, 1931) in all Reno motel rooms. The least sober of gamblers, on reading the precise formulation and proof that no *system* can improve the bettor's fortunes, will instantly repack, check out, and rush back to his or her loved ones. Some may possibly return to their spouses and families.

**Voltaire-Candide, law of** "All is for the best in the best of all possible environments." (originally: "Tout est pour le mieux dans le meilleur des mondes possibles."—*Candide*, Voltaire.)

♦ A cynical 18th-century acceptance of the status quo adopted by computer users in the 20th century, but not without some envy of the relatively trouble-free adventures enjoyed by Candide and Pangloss. Among the many familiar observations supporting the law, we offer:

"God sent us this 360, and Lo! our 1400 payroll programs run no slower than before."

"The six-month delivery setback will allow us to refine our flowcharts and build a computer room."

"The file I have just accidentally erased was due for purging sooner or later."

"The more data I punch in this card, the lighter it becomes, and the lower the mailing cost."

"Our system has broken down. We can all retire to the canteen, where the on-site engineer is watching the Big Fight on TV."

"This flowchart, although rejected in toto by the DPM, will nicely cover the crack in the wall above my desk."

"The system has crashed just as I was beginning to suspect an endless loop situation."

"We were freezing during the power outage, until the standby generator caught fire."

# The generic word processor

## A word-processing system for all your needs†

*Philip Schrodt*

Congratulations on your purchase of the GWP Inc. Generic Word Processor System. We are sure that you will find this word processor to be one of the most flexible and convenient on the market, as it combines high unit reliability with low operating costs and ease of maintenance.

Before implementing the system, carefully study figure 1 to familiarize yourself with the main features of the GWP word-processing unit.

**CHARACTER DELETION SUBUNIT**

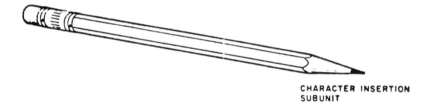

**CHARACTER INSERTION SUBUNIT**

**Figure 1.** The GWP System word-processing unit is composed of the character-insertion subunit (at right) and the character-deletion subunit (at left).

## Initialization

The word-processing units supplied with your GWP are factory-fresh and uninitialized. Before they can be used, they must be initialized using the GWP initialization unit

†**Editor's Note:** Once in a while we come across a product that's so useful we feel compelled to bring it to our readers' attention. The Generic Word Processor System (GWP) is such a product, incorporating the essentials of a word processor in a sublimely simple form.

With the manufacturer's permission we are reprinting the documentation for this product. After working with the GWP for several weeks, we're delighted by the feeling of total control that the system gives us and are certain you will be too. No more accidentally erased files, no damaged disks, no hardware problems ... SJW.

(see figure 2). Because of the importance of this unit, we designed it with a distinctive shape so that it will not be misplaced among the voluminous vital papers on your desk.

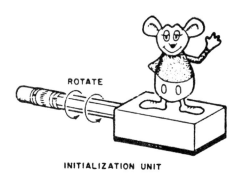

ROTATE

INITIALIZATION UNIT

**Figure 2.** The word-processing initialization unit, which should be operated over a wastebasket.

To initialize a word-processing unit, carefully place the character insertion subunit into the left side of the initialization unit and rotate the word-processing unit approximately 2000 degrees clockwise while exerting moderate pressure on the word-processing unit in the direction of the initializer. Check for successful initialization by attempting a character insertion. If the insertion fails, repeat the initialization procedure. The word-processing unit will have to be reinitialized periodically; do this whenever necessary. (*Warning:* do not attempt to initialize the word-processing unit past its character deletion subunit. Doing so may damage both the word processor and the initializer.)

### Operating the Word Processor
The GWP can perform all the basic functions featured in word processors that cost thousands of dollars more.

Furthermore, because the GWP does not require electricity, it can operate during power blackouts, electrical storms, and nuclear attacks. By conserving precious energy resources, it helps free our beloved country from the maniacal clutches of OPEC.

Basic functions of the word processor are listed below:

*Inserting text:* Use the character-insertion subunit to write in the words you wish to insert, applying moderate downward pressure to the unit. Be sure to write clearly so that the typist can follow what you have written.

*Deleting text:* With moderate downward pressure, rub the character-deletion subunit across the text to be deleted. Repeat this procedure several times. The text will gradually disappear, whereupon you will be able to insert new text.

*Underlining:* Using the character-insertion subunit, place the unit slightly below and to the left of the first character you wish to underline. Move the unit to the right until you reach the last character to be underlined.

*Bold face:* Repeat the text-insertion procedure twice, pressing downward with greater pressure than you would normally apply.

*Move to beginning of text:* With the text you are working on in hand, move the unit to the beginning of the text.

*Move to end of text:* Take the text you are working on and move the unit to the end of the text.

*Moving blocks of text:* Block moves require use of the block text extraction unit and the block text replacement unit pictured in figure 3. By means of the block text extractor unit, sever the paper immediately above and below the text you wish to move. Instructions for operating the extractor unit are etched on the side of the unit in Korean. If you still have difficulty operating the unit, call our service department for consulting help at our introductory fee of $50 per hour, or ask any 5-year-old child.

After separating the text to be moved, open the lid of the block text replacement unit and, grasping the block text replacement medium application unit, spread the block text replacement medium on the back of the text. Move

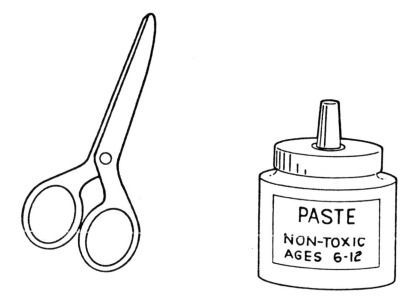

**Figure 3.** The block text extraction and replacement units, commonly run in unison.

the text to the new location and affix it to another sheet of paper with gentle but firm pressure. In a few minutes, your text will be permanently affixed in the new location.

### Other Features

*Page numbering:* After writing your entire text, inscribe a 1 on the first page, a 2 on the second page, etc. When you finish, all of the pages will be numbered.

*Centering:* Determine where the center of the page is by looking at it. The center is usually near the middle of the page. Place the text to be centered evenly on each side of the center. It is now centered.

*Special fonts:* The GWP System is extremely versatile and easily adaptable to specialized type fonts such as Sanskrit, Amharic, and hieroglyphics (see figure 4). You will find these fonts valuable in business correspondence, particularly if you are in frequent contact with Vedic gurus, Egyptologists, or Ethiopian Airlines.

*Saving files:* Put the work you have finished in a safe

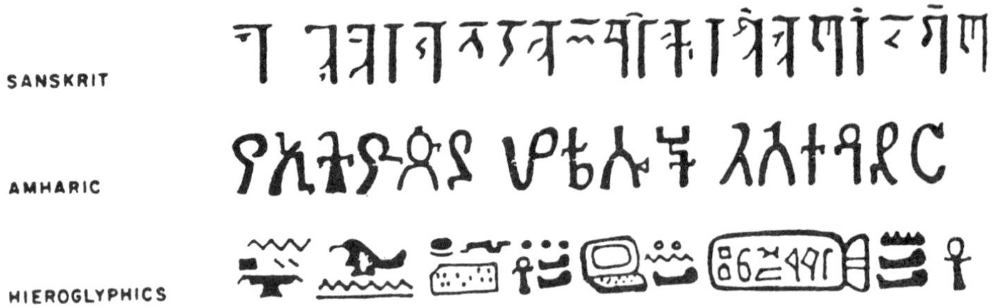

**SANSKRIT**

**AMHARIC**

**HIEROGLYPHICS**

**Figure 4.** Sample type fonts illustrating the wide variety available with the GWP System. The manufacturer claims that if a language can be written, the GWP System can be adapted to it.

place, one where nobody will find it or spill coffee on it. If it is not disturbed, it will be there when you return.

*Deleting files:* Take any files you no longer need and deposit them in the wastebasket. They probably will be gone in the morning. In most offices, this can also be accomplished by leaving the files in the open, forgetting to remove them from the copying machine, or writing CONFIDENTIAL on the file in bold letters.

*Appending files:* Place the first file on top of the second file. Treat the two files as though they were one file.

*Justification:* Most word processors have little justification. This word processor has no justification at all, as it does not even lend prestige to the office where it is used, which is the justification for most word processors.

### Printing Files

A printer for the GWP must be purchased separately. For convenience of operation, we recommend an ordinary typewriter and a typist. Give the text to the typist and tell him or her to type it. Printing speed can be improved by increasing the wages of the typist, threatening to withhold the wages of the typist, kidnapping pets, plants, or children of the typist, instigating intimidating tactics, and other conventional office-personnel management techniques. Printing speed can be decreased by asking to see the text, making continual changes in the text, asking the typist to answer the phone, decreasing the typist's wages, and installing a conventional electronic word processor. You will soon learn to adjust the printing speed to the optimal level for your particular needs.

Copyright © 1981, Generic Word Processing Inc., Skokie, IL 60076

---

# Computing machinery and intelligence

*A M Turing*

A M Turing
*MIND* **59** No 236,
Oct 1950, p 433.
RBP from Oxford
University Press.

[In one of the famous early papers on computers, Turing raised the question, 'Can they think?' He presented definitions and arguments from nine viewpoints, theological to mathematical. The following excerpt from his 24-page paper presents Turing's personal critique of the problem.]

## 1. *The Imitation Game*

I propose to consider the question, 'Can machines think?' This should begin with definitions of the meaning of the terms 'machine' and 'think.' The definitions might be framed so as to reflect so far as possible the normal use of the words, but this attitude is dangerous. If the meaning of the words 'machine' and 'think' are to be found by examining how they are commonly used it is difficult to escape the conclusion that the meaning and the answer to the question, 'Can machines think?' is to be sought in a statistical survey such as a Gallup poll. But this is absurd. Instead of attempting such a definition I shall replace the question by another, which is closely related to it and is expressed in relatively unambiguous words.

The new form of the problem can be described in terms of a game which we call the 'imitation game.' It is played with three people, a man (A), a woman (B), and an interrogator (C) who may be of either sex. The interrogator stays in a room apart from the other two. The object of the game for the interrogator is to determine which of the other two is the man and which is the woman. He knows them by labels X and Y, and at the end of the game he says either 'X is A and Y is B' or 'X is B and Y is A.' The interrogator is allowed to put questions to A and B thus:

C: Will X please tell me the length of his or her hair?

Now suppose X is actually A, then A must answer. It is A's object in the game to try and cause C to make the wrong identification. His answer might therefore be

'My hair is shingled, and the longest strands are about nine inches long.'

In order that tones of voice may not help the interrogator the answers should be written, or better still, typewritten. The ideal arrangement is to have a teleprinter communicating between the two rooms. Alternatively the questions and answers can be repeated by an intermediary. The object of the game for the third player (B) is to help the interrogator. The best strategy for her is probably to give truthful answers. She can add such things as 'I am the woman, don't listen to him!' to her answers, but it will avail nothing as the man can make similar remarks.

We now ask the question, 'What will happen when a machine takes the part of A in this game?' Will the interrogator decide wrongly as often when the game is played like this as he does when the game is played between a man and a woman? These questions replace our original 'Can machines think?'

70

## 2. *Critique of the new problem*

As well as asking 'What is the answer to this new form of the question,' one might ask, 'Is this new question a worthy one to investigate?' This latter question we investigate without further ado, thereby cutting short an infinite regress.

The new problem has the advantage of drawing a fairly sharp line between the physical and the intellectual capacities of a man. No engineer or chemist claims to be able to produce a material which is indistinguishable from the human skin. It is possible that at some time this might be done, but even supposing this invention available we should feel that there was little point in trying to make a 'thinking machine' more human by dressing it up in such artificial flesh. The form in which we have set the problem reflects this fact in the condition which prevents the interrogator from seeing or touching the other competitors, or hearing their voices. Some other advantages of the proposed criterion may be shown up by specimen questions and answers. Thus:

*Q:* Please write me a sonnet on the subject of the Forth Bridge.

*A:* Count me out on this one. I never could write poetry.

*Q:* Add 34957 and 70764.

*A:* (Pause about 30 seconds and then give the answer) 105621.

*Q:* Do you play chess?

*A:* Yes.

*Q:* I have K at my K1, and no other pieces. You have only K at K6 and R at R1. It is your move. What do you play?

*A:* (After a pause of 15 seconds) R – R8 mate.

The question and answer method seems to be suitable for introducing almost any one of the fields of human endeavour that we wish to include. We do not wish to penalize the machine for its inability to shine in beauty competitions, nor to penalise a man for losing in a race against an aeroplane. The conditions of our game make these disabilities irrelevant. The 'witnesses' can brag, if they consider it advisable, as much as they please about their charms, strength or heroism, but the interrogator cannot demand practical demonstrations.

The game may perhaps be criticised on the grounds that the odds are weighted too heavily against the machine. If the man were to try and pretend to be the machine he would clearly make a very poor showing. He would be given away at once by slowness and inaccuracy in arithmetic. May not machines carry out something which ought to be described as thinking but which is very different from what a man does? The objection is a very strong one, but at least we can say that if, nevertheless, a machine can be constructed to play the imitation game satisfactorily, we need not be troubled by this objection.

It might be urged that when playing the 'imitation game' the best strategy for the machine may possibly be something other than imitation of the behaviour of a man. This may be, but I think it unlikely that there is any great effect of this kind. In any case there is no intention to investigate here the theory of the game, and it will be assumed that the best strategy is to try to provide answers that would normally be given by a man.

...

We may hope that machines will eventually compete with men in all purely intellectual fields. But which are the best ones to start with? Even this is a difficult decision. Many people think that a very abstract activity, like the playing of chess, would be best. It can also be maintained that it is best to provide the machine with the best organs that money can buy, and then teach it to understand and speak English. This process could follow the normal teaching of a child. Things would be pointed out and named, etc. Again, I do not know what the right answer is, but I think both approaches should be tried.

We can only see a short distance ahead, but we can see plenty there that needs to be done.

['There are some who, confronted with this bewildering maze of developments, wonder whether the first such electronic brain, when created, will promptly turn off the nearest television set and disconnect all similar appliances in its area, and settle down for a quiet evening with a good old-fashioned book.'—Maurice B Mitchell in 'A Forward Look at Communications' 1958, Britannica Book of the Year.]

THE WAY I SEE IT, IT DOESN'T MUCH MATTER WHICH THEORY YOU BELIEVE JUST SO LONG AS YOU'RE SINCERE.

AND I SAY "DOWN WITH THE EXCLUSION PRINCIPLE— STRIKE A BLOW FOR FREE ELECTRONS!"

# Classroom Foibles and Folies

The light in the world comes principally from two sources—the sun, and the student's lamp.

*Bovee*
1842

# What every quantum mechanic should know

*Kurt Gottfried*

[Kurt Gottfriend teaches second semester graduate-level quantum mechanics at Cornell University. The following quotes come straight from his lectures.]

Contributed by Kristin S Ralls

*About pedagogy:*
- Part of the art of teaching is not to tell the truth completely.
- This subject is no longer suitable for an examination ... of the usual type.
- Studying for the exam won't help you anyway.
- There's a lot of deep significance to that. I'm just telling you. Just to mystify you.
- If you fall asleep I'm sorry but it has to be done.
- Now that we're out of time I want to cover one more thing.

*About calculations, and especially how long they take:*
- If you do everything properly and come back a year later you'll find it's still divergent—it's logarithmically divergent.
- Can I do perturbation theory, or do I have to go away for six years to learn how to do such a calculation?
- You can go on for the rest of your life now, looking at the rest of the excited states.
- It's quite clear that the two-body density matrix is an unpleasant thing to calculate ... for that reason I will let you do it yourself.
- A certain measure of prudence is necessary...
- At some point you have to know about the wave functions; you can't do without them forever... So, we have to calculate this miserable thing, and it gets to be a little bit miserable.
- It's trivial ... it's non-trivial ... it requires calculation ... *somebody* has to do it! *Once!* Like Mr. Fock.

*About errors and mistakes:*
- If you find a partial wave cross-section larger than $4\pi\lambda^2(2l + 1)$, you have either made a mistake or

shown that quantum mechanics is invalid.
- So if I've made no mistake this is the answer.
- This is correct except for some errors.
- As long as you don't make a mistake, you can do whatever you like.

*About formulas and notation:*

- $\langle 1'' | U_{\text{dir}} | 1 \rangle = \delta(1''1')z \int v(r_1' - r_2)\langle 2| \rho |2 \rangle d2$

  Now this can actually be understood by a human being... This is obviously what any child would have written down... It's just an old-fashioned potential, the kind you heard about—you know, wherever you did your undergraduate work.
- It's a standard thing in physics to simplify things by using more and more obscure notation.
- You develop a notation so that anyone who walks into the room after that point can't understand what's going on. Of course it's a very powerful notation for those who are members of the cult.
- I will now, for reasons that will become obvious in a minute, write this nice formula in a more complicated way.
- Dirac of course, as always, has invented the world's most beautiful formalism (it really is)... and you don't have to think!
- Dirac notation is the best notation in the history of physics. I'm serious—if you take a course on general relativity you quickly discover that.
- This is the usual Dirac mumbo-jumbo...
- Why didn't I put + and −? Well of course you know somebody wrote a paper thirty-five years ago and that's the way we all do it nowadays.

*About other physicists:*
- People with two PhD's get this wrong all the time ... *not* Cornell people!
- I know from experience that there's an amazing number of professionals who do not understand this...

*About Landau:*
- ... who was not the first person to do this, and who I think was always furious that he wasn't.

*About symmetry breaking:*
- The state does not have the full symmetry of the Hamiltonian ... unless you've gone to a good graduate school you won't know the Hamiltonian is symmetric.

*About simplifying hard problems:*
- In a way what you do is change the rules of the game until you like them (being careful to avoid throwing the baby out with the bath water).
- We have *not* used perturbation theory—we have used an axe on the Hamiltonian.
- Well, the proof is not difficult... We've already done the hard work of getting ourselves to this childish Hamiltonian...
- Now the whole thing has become child's play ... *that's* the kind of integral everybody loves.
- This is an exact formula—I've done nothing except fiddle around ... of course if you want to calculate this thing you have to do a lot of work.
- If the basketball is filled with basketballs this doesn't work.

*About computers:*
- You don't know anything, but you have a big computer at your disposal—this is getting to be a real disease.

*About theorists:*
- (To get higher order terms) you'd have to get somebody from the fifth floor of Clark Hall, and he wouldn't do it this way anyway.

*About rules of thumb:*
- 1 is not interesting—nothing is happening. That's the difference between nothing happening and something happening.
- Particle 3 is bald and particle 4 has four legs...

- Even in quantum mechanics you can't divide by zero.
- That's almost always the case, except when it's not.
- The exchange interaction cares *deeply* what the spin structure is.
- You have to start someplace, even in physics.
- … quantum mechanics kills you.

*About angular momentum technology:*
- … a marvelous superstructure of Baroque complexity.
- You can spend a couple of days learning about Racah coefficients, and then there's nothing to do … in this case it's better to be a pedestrian.

*About famous results:*
- … there's some miserable formula, which no one can remember.
- You may or may not have known that but it's still true.

# A problem in quantum mechanics

*Norman Ramsay*

[This is a quantum mechanics problem suggested by Norman Ramsay when a first year graduate student at Cornell University.]

Contributed by Kristin S Ralls.

Consider the operator $a_g^+$, which awakens a graduate student.

A graduate student is enjoying an eigenstate of the sleep operator $s = a_g^+ a_g$ when, at 7:00 a.m., a ringing alarm clock is applied to the system. What is the perturbation Hamiltonian? What is the time evolution of the wave function? The same student, now in a coherent superposition state, goes to an 8:30 class. What is the probability that he (or she) will remain awake in class? Quantum mechanical interference effects are particularly pronounced in this system. Show that the usual effect causes the student to fall asleep every minute or two, and then to awake w/a start (+ a guilty look). EXTRA CREDIT: Using this formalism, write the time-dependent coffee operator.

# Near misses

Gordon Hanson, De Havilland College, Welwyn Garden City, Herts., UK

Here are some recent examination paper howlers from my own students. I always find these slightly depressing as well as humorous.

1. When under excessive strain, copper becomes plasticene.
2. At the dew point, air turns into water.
3. Radio waves are able to encircle the globe by bouncing off the heavy layer which surrounds it.
4. A ballistic or a periodic galvanometer is similar in most respects to an ordinary one except that it is undamned.
5. *Self:* We have milli-ohms and meg-ohms in electronics, what other prefixes are used?
   *Class:* Sherlock ohms, home sweet ohm.
6. *Self:* What would you say that platinum-based thermocouples have in common with science teachers? [Expected answer was noble, incorruptible, etc.]
   *Class:* Inert, sir?
7. Brownian movement was invented by Robert Browning.
8. The law of conservation of energy cannot be destroyed.
9. Charge can be measured by using a majestic galvanometer.

---

# Science fiction

*Anon*

Contributed by Mary Gail K Hutchins.

Do you recall when you were first struggling to master the mysteries of Science, vainly trying to remember the formula for table salt, or whether the guy who ran around shouting 'Eureka!' was Archimedes or Aristotle?

Perhaps, during those dark and confusing days, you contributed a line like one of the following—all of which are taken from the papers of fledgling Science students.

- A thermometer is an instrument for raising temperance.
- There are three kinds of blood vessels: arteris, vanis, and caterpillars.
- A magnet is something you find in a bad apple.
- To remove air from a flask, fill the flask with water, tip the water out, and put the cork in quick.
- A litre is a nest of young baby animals.
- The earth makes a resolution every 24 hours.
- Typhoid fever may be prevented by fascination.
- Algebrical symbols are used when you do not know what you are talking about.
- Geometry teaches us to bisex angels.
- A flower's only protection against insects is its pistol.
- An example of animal breeding is the farmer who mated a bull of good meat to a bull that gave a great deal of milk.
- If conditions are not favorable, bacteria may go into a period of adolescence.
- We believe that the reptiles came from the amphibians by spontaneous generation and the study of rocks.
- By self-pollination, a farmer may get a flock of long-haired sheep.
- A triangle which has an angle of 135 degrees is called an obscene triangle.
- When you haven't got enough iodine in your blood, you get a glacier.
- It's a well-known fact that a decreased body warps the mind.
- The hydra gets its food by descending on its prey and pushing it into its mouth with its testacles.
- C H2O is the formula for seawater.
- Algebra was the wife of Euclid.

What's worrisome is that some of those sound pretty reasonable to me. Not the last one, though. Everybody knows that Euclid's wife was Polly Hedron.

# Advantages of physics

Charles Dyer in
*The Physics
Teacher*
September 1968
p 321. RBP from
*The Physics
Teacher*.

*Physics is the most basic of sciences.*
*What other science can describe the trajectory*
*   of a baseball in a vacuum, or needs to?*
*In Experiment One we discovered the latent properties*
*   of the meter stick.*
*Verification of gravity was next with the discovery of*
*   friction in Part B, figure 12-12.*
*In a year or two we will discover over again*
*   the ratio of charge to mass of an electron!*
*(In case anyone forgot.)*

*Physics is not good for everyone though. This is especially*
*   the case for children of those who are over-exposed*
*   to radiation.*
*But few mind the hardships, as positions in physics*
*   command great respect, since the bomb.*
*With popular support and protection from the*
*   government, how could a guy miss?*

---

# Outline of recent advances in science

[Specially designed for members of women's culture clubs, and representing exactly the quantity of information carried away from lectures on scientific progress]

From *Winnowed
Wisdom A New
Book of Humour*
by Stephen
Leacock 1926
(New York:
Dodd, Mead &
Co) pp 25–28

*Einstein's Theory of Relativity*
Einstein himself is not what one would call a handsome man. When seen by members of the Fortnightly Women's Scientific Society in Boston he was pronounced by many of them to be quite insignificant in appearance. Some thought, however, that he had a certain air of distinction, something which they found it hard to explain but which they *felt*. It is certain that Einstein knows nothing of dress. His clothes appear as if taken out of the rag bag, and it is reported by two ladies who heard him speak at the University of Pennsylvania on the measurement of rays of light that he wore an absolutely atrocious red tie. It is declared to be a matter of wonder that no one has ever told

82

him; and it is suggested that someone ought to take hold of him.

Einstein is not married. It has been reported, by members of the Trenton (New Jersey) Five O'clock Astronomical Investigation Club that there is a romance in his life. He is thought to have been thrown over by a girl who had a lot of money when he was a poor student, and it was this that turned his mind to physics. It is held that things work that way. Whether married or not he certainly behaved himself like a perfect gentleman at all the clubs where he spoke. He drinks nothing but black coffee.

Einstein's theories seem to have made a great stir.

******

*Madame Curie's Discoveries in Radio-Activity*
Madame Curie may be a great scientist but it is doubted whether she is a likeable woman or a woman who could make a home. Two members of the Omaha Woman's Astronomical and Physical Afternoon Tea Society heard her when she spoke in Washington on the Radiation of Gamma Particles from Helium. They say that they had some difficulty in following her. They say she was wearing just a plain coat and skirt but had quite a good French blouse which certainly had style to it. But they think that she lacks charm.

******

*Rutherford's Researches in the Atomic Theory*
Ernest Rutherford, or rather Sir Ernest Rutherford as it is right to call him because he was made a knight a few years ago for something he did with molecules, is a strikingly handsome man in early middle age. Some people might consider him as beginning to get old but that depends on the point of view. If you consider a man of fifty an old man then Sir Ernest is old. But the assertion is made by many members of various societies that in their opinion a man is at his *best* at fifty. Members who take that point of view would be interested in Rutherford. He has eyes of just that pale steely blue which suggest to members something

powerful and strong, though members are unable to name it. Certainly he made a perfectly wonderful impression on The Ladies Chemico-Physical Research and Amusement Society in Toronto, when he was there with that large British body.

Members of clubs meeting Sir Ernest should remember that he won the Nobel Prize and that it is not awarded for character but is spelled differently.

---

# Letters to a chemistry professor

*R T Sanderson*

RBP from the *Journal of Chemical Education* 1978 **55** 454–55

Although the Official World deems me old enough to retire, I shall never get over my childish excitement over the daily mail. It is like fishing in an unknown pool. One never knows what will be on the line he pulls up, and the anticipation lends zest that is renewed each day. One can never feel imprisoned in his ivory tower when he knows there are thousands of others 'out there' and that some of them sooner or later may wish to communicate in a personal way.

Among the most common types of personal mail, a chemistry professor may distinguish four principal categories: reprint requests, scientific discussions, requests for information, and letters of appreciation. I should like to comment briefly on each of these, in that order.

Reprint requests, which from civilized countries having an abundance of copying machines have diminished greatly in recent years, may be regarded as nuisance by some, but to me they are a welcome indicator of the extent of interest. When, as has been my common experience, the same paper is requested by botanists, chemical theorists, geologists, biochemists, physicists, agronomists, nuclear scientists, foresters, physical chemists, and pharmacologists among others, one gains needed reassurance that his ivory-towered researches may have some useful general significance, a practical illustration of the truly fundamental importance of chemistry. My favorite people are the ones who take the trouble to insert on the formal printed request, 'very interesting,' or make similar comments in a personal note. A little farther down on my list of favorites are the folks who request reprints of all one's previous work, which over a period of forty years can amount to a fairly large order, and of course practically impossible to fill. Most touching are the requests from impoverished countries for copies of one's books. Unless the author is very wealthy, he must be fairly critical of these requests, for they can only be filled at substantial personal expense. Contrary to common belief, publishers do not give authors an infinite supply of free copies of their works.

The second category, 'scientific discussions,' includes the disclosure of errors in the author's published writings. Discussions of one's research are of course always welcome, for they can be very helpful, challenging, and stimulating. As for publishing, it seems almost impossible to print a book without errors. Glaringly obvious typographical errors can sneak by several proof readers in a row, just because each knows in advance what the word is supposed to be and cannot recognize that it is garbled in print. Then as soon as the book is published, these same errors leap out like naughty children hidden under a honeymoon bed. But they are not nearly as serious, however annoying, as mistakes in fact. The author has a responsibility to be omniscient, because people reading a textbook expect it to be true. The expectation of omniscience is of course completely unjustified, because even a spotless author is at the mercy of mistakes printed by the authors from whom he copies.

A good case can be made for deliberately including at least one or two factual mistakes in every textbook. This promises a wonderful feeling to the reader who finds and recognizes these mistakes. 'Even I know better than that!' he exults, feeling on top of the world. If the reader is a teacher, what enormous pleasure he can derive from pointing out these mistakes to his students! 'Did you see that statement at the bottom of page 29? That just isn't true. The author really goofed there!' What student could fail to admire a teacher who knows more than the author of his textbook? And if the reader is a student, how exhilarating the feeling of discovery! 'Hey, man, did you read what the book says on page 29? That's wrong, man! I looked it up in two other books, and the truth is just the opposite. Of course I just checked to make certain—I knew something was wrong right away. Hey, teach, this book is wrong!'

If the reader is outside the classroom, he may lack an appreciative audience to whom he can brag that he has found a mistake in the book. 'Hey, look here,' he says to his wife, 'this author writes that water is a compound of nitrogen. Whatever was he thinking of? There's no nitrogen in water.' And his wife says, 'That's nice, dear, the beans are boiling over,' and away she runs. That leaves him with the option of writing to the author. If he waits a few days he will probably never get around to writing that letter. But if he writes before the white heat of excitement has cooled, he may even receive a thankful acknowledgment from the author. Then he can nonchalantly tell his friends, 'Yeah, I know that book, I found a couple of mistakes that I pointed out to the author, and he was glad to know about them.'

Sometimes the letter to the author will be very polite.

Dear Dr Blank: In reading your very excellent new book recently, I encountered the statement that water contains nitrogen. This does not seem consistent with the chemical formula, $H_2O$, which I had always been told is the accepted formula. Therefore I wonder if you would please be so kind as to give me a reference to your source of information concerning the nitrogen content of water. Respectfully yours.

Or, it may read as follows

I think you may wish to have your attention directed to what seems to be an unintentional inadvertency in your otherwise very admirable textbook of recent publication date. May I quote from the middle of page 326 of this book: 'Water is a compound containing nitrogen and oxygen atoms chemically bonded together.' It is my personal belief that someone reading this book might get the impression that one of the elements in water is nitrogen. I do not believe it was your intention to give such an impression, so if I should happen to be correct, perhaps you will wish to suggest a small change for your next printing.

Sometimes there may be less tolerance of the author's lack of perfection.

Dear Dr Blank: How in the world can we expect our students to go forth with a reasonable understanding of chemistry if they read in a book such as yours a ridiculously erroneous statement (page 326) that water is a compound of nitrogen. I feel impelled to tell you that not only will we not adopt your book as a textbook for our students, but also I have issued orders that no copy is to be purchased for the library or bookstore. Indignantly yours.

Any author should be embarrassed to have an error pointed out, but he should nonetheless be grateful, so that a correction can be made at the earliest opportunity. Please do help the author, gentle reader, for it cannot be taken for granted that someone else will have pointed out each mistake. Many years ago I was shocked to observe that I had forgotten a methyl group on a molecular model on display, for that same model had been photographed and the photo published in a book. For 15 years I have cowered in fear and trembling behind retired garments in a dark attic closet, awaiting the inevitable avalanche of scornful, scolding letters. So far, not a single soul has told me of this error. Now I am hoping it never will be discovered, and lately have even been coming out at night for brief, solitary walks.

Requests for information form the third principal category. Some of these are fascinating for their imagination and scope. Once came a card from a high school teacher in South Dakota. 'Dear Sir: At an NSF institute in Ohio last summer we were told that you are an expert on chemical bonds. Please send me information on chemical bonds.'

More commonly, letters come from students with assignments. Either they are looking for a project for a science fair, or they have a special term paper to write. For example, here is a message from a Puerto Rico high school boy: 'I have an assignment to write about chlorine. Please send me all the information, brochures, booklets, and books you have about chlorine. Please rush as my assignment is due next Friday.' And this from a boy in Virginia: 'For my science fair project, please send me samples of each of the chemical elements. Also samples of any compounds of any of the elements which you might have. Yours truly.' A hopeful young Texan wrote, 'My teacher said you make models. I think your idea is a good one. Please send me materials to make models. Also colors to color them with. Also whatever else you think would be handy for me to have.'

One of my favorite correspondences was a brief one with a young lady in Massachusetts. She wrote for instructions on making chemical models, a list of which she provided, to help her in a science fair project. I sent her detailed instructions covering her complete request. No further word came for several months. Then one day an airmail special delivery letter: 'My project of models has just been awarded first prize in our local science fair. It will now be entered in the finals of the statewide competition. Please write me at once and tell me what everything stands for.'

Now the fourth and final category, the one you have all been waiting for, the letters of appreciation. I truly hope you will not be too disappointed, but I am entirely too modest to reveal the content of those. However, I will make two exceptions, to give you an idea of them.

One is the following

> I am just a chemistry student but I must tell you how much I appreciate your fine new book. It is of enormous help to me in learning chemistry and gives me insights and understanding that I never thought would be possible for me to have in view of the great difficulty of the subject which is much different than I had in high school. So, Professor Sanderson, I am forever in your debt for your having taken the great trouble to explain this chemistry so simply and clearly that even I could understand it. This is my most valuable book. Thank you again for writing it.

I have only read the first page but am looking forward with great excitement to reading the rest of it. Yours truly,

The second example was received from a beautiful pink-cheeked blond, a third-grader to whose class I had talked about chemistry. 'Dear Dr Sanderson, Before you came, I hate chemistry. Now it is my favorit. I am going to Grandma's for Thanksgiving. Love, Peggy.'

# The ergodic skeleton in the cupboard of thermodynamics

*M V Vinor*

RBP of Professor A Kohn from *Journal of Irreproducible Results* **6** 11 (1958)

*The Royal Road of Thermodynamics* is considered to be the best one for leading the student towards the *Realm of Higher Sophistication*. As such it presents to the novice a magnificent front of awe-inspiring Laws and sweeping generalizations which seem as well founded and unshakable as the *Scientific Method* itself. But this is only true as far as the superstructure itself is concerned. At the very foundations the Devil is lurking in the form of a *Demon Assumption* known as the *Ergodic Hypothesis*.

The student of thermodynamics is not likely to find this hypothesis even mentioned in passing in his usual textbooks because the authors are afraid of spoiling his faith in the subject; they contrive to put it in when he is not looking.

So we find ourselves compelled to make public this Unmentionable Item and thus acquire *Scientific Credit* at the expense of everybody else concerned. We are positively not going to be sorry if the student, too, happens to benefit in the process. We only advise him not to utter these *Heretic Opinions* until he has received his Degree.

We know, from everyday experience, that our world is not ideal. Not all girls are beautiful, rich and in love with us, etc. When we told our teacher that a swing does not swing forever, as it should in Physics, and we had to work hard to keep our girl-friend (on the swing) in motion, he told us that because of 'mechanical friction' energy is converted into Heat, but ideally … and so on. The same holds when a battery is discharged; we always pay something to Heat because of 'electronic friction'. When we came to Thermodynamics we encountered the same process with Heat itself: two bodies at different temperatures settle down to equilibrium instead of oscillating forever as a pendulum or electrically charged spheres should (except for that damn 'friction'). Being smart guys, we asked: 'Where and what kind is Heat Friction?' Bitter personal experience leads us to advise the student not to ask this question, because it touches IT.

Molecules in a gas are assumed to be mechanically perfect and to undergo elastic collisions (note: no friction!). Have we here an ideal system? Not at all! When a gas expands to fill an evacuated container it comes to an equilibrium of Maxwellian distribution fairly quickly and apparently it does not expend any energy. This begs a question. If there is no 'friction' it should go on oscillating forever. The simple observation that it does not, goes under the cover name of 'ERGODIC HYPOTHESIS or ASSUMPTION' and can be otherwise stated as follows:

'*Ergodic Assumption:* In every beautiful theory there is always a catch.'

Since this formulation may not be too clear to the layman, scientists prefer to say (when they are compelled to say anything about it) that… 'The system is ergodic and therefore…' Teachers and lecturers usually say: '…and at equilibrium we have…' without bothering to tell the student that in order to reach an equilibrium some sort of 'heat friction' is required; they would never admit that it existed. Many have never heard the adjective *ergodic* and these mislead their students with a clear conscience. Those who have, hardly ever dare to show it.

# Gems from the years, or To err is human

RBP of the publisher, 'Gems from the years, or To err is human', *CMAJ* **113**, October 1975 p 616.

Examination boners or howlers, and indiscretions of spelling have been presented in the past but each generation of students manages to mint a few originals and, if the metaphor is not too mixed, among them an occasional masterpiece. Such miscarriages of language are likely as old as writing itself, but surely some of those to follow rank with the best of any age.

The source of the gems has been the examination papers of medical and nonmedical students, graduate and undergraduate, foreign and homegrown, all at the University of Toronto. A few introductory examples are offered to set the spirit of the piece and to encourage readers to continue—the reward being a smile or two along the way.

For starters there is the learned and original remark that 'people who are carriers should be confiscated'. This suggests to me that the perpetrator is badly in need of a vocabulary review. Another culprit also has it in for carriers and states that 'carriers may well be the up-starts of devastating and widespread epidemics'. An arrogant lot, these carriers. Lastly, as an introduction an examinee has this to say: 'Disease and death are the cause of many fatalities.' A morbid outlook, surely.

Now, professors like to organize and classify their material, so perhaps at the head of the list a few bits of 'wit and wisdom' relating to venereal disease are in order. A few of the better examples are hereby recorded together with some humble comments.

'Gonococcus is a venereal disease organism stained with genital violet.' [Sounds more apropos then gentian.]

'*Treponema pallidum* from sexuous exudate.' [Good try.]

'The best prevention for syphilis is to go home after drinking.' [An old head on young shoulders?]

'Droplet infection refers to sexual contact.' [Oh, yes?]

'*Treponema pallidum* is a yeast.' [Would that it were.]

'*Treponema pallidum* are certain bacteria which appear in palisade layers—an example is *C. diphtheriae*.' [Confusion reigneth supreme.]

'If Henry VIII hadn't had syphilis he wouldn't have divorced Katherine, wouldn't have got in a quarrel with the Pope and England might be a Catholic country to-day. day.' [A historian at heart.]

And a few related misspellings: 'shanker', 'Spyrockaetes', 'Spyrokytae'. [They're bad words, anyway.]

BCG came in for some remarkable suggestions. One budding homespun philosopher proposed that 'BCG (Bacille-Calmette-Guerin—a tuberculosis vaccine) is used on Indians, doctors, nurses and slums.' [Strikingly put, and not too wide of a bull's eye at that.] A few additional descriptions of BCG follow.

'—three Frenchmen.' [Overdoing it a bit.]

'—Bovine-Calmette-Guerin'. [Closer.]

'...massaged over thirteen years.' [Try 'passaged'.]

'It is a bovine type typhoid bacillus ... and is named after the three men who thought it up.'

'...is tested in the giny pig.' [A phoneticist apparently.]

'BCG is used to fight against polio which causes smallpox.' [Pretty good stuff by the sound of it.]

An unusual sign of tuberculosis was also proposed: 'Look for emancipation in the animal.' [Emaciation would be less surprising.]

# Bureaucracy

The reasonable man adapts himself to the world; the unreasonable one persists in trying to adapt the world to himself. Therefore all progress depends on the unreasonable man.

*George Bernard Shaw*

# 6 phases of a project

'Office humour' offered by David Broome.

1. **Enthusiasm**
2. **Disillusionment**
3. **Panic**
4. **Search for the Guilty**
5. **Punishment of the Innocent**
6. **Praise and Honors for the Non Participants**

---

## A minimal/maximal handbook for tourists in a classified bureaucracy

*Amrom H Katz*

Condensed from 'A Guide for the Perplexed' by Amrom H Katz Air Force/Space Digest, November 1967. RBP of the Air Force Association, Washington, DC.

Some fifteen years ago, after more than ten years of life as an entrenched bureaucrat and not-too-civil servant behind the lines in the Air Force's Aerial Reconnaissance Laboratory at Wright Field at Dayton, Ohio, I was assigned, and was able to successfully carry off, the role of guide, adviser, and philosopher to a group of younger associates.

I had been struck over the years by the continual recurrence in the jungles of bureaucracy of patterns of conflict, concealment, collusion, confusion, and corruption. Thus, when I was faced with the problem of providing guidance and insight for younger, softer-skinned, and wondering associates (all of whom had had less experience and success than I in fighting city hall), I found it necessary to produce a set of maxims which wrap up most of the more obnoxious aspects of life in a bureaucracy.

**MAXIM 1:** Where are the calculations that go with the calculated risk?

**MAXIM 2:** Inventing is easy for staff outfits. Stating a problem is much harder. Instead of stating problems people like to pass out half-accurate statements together with half-available solutions which they can't finish and which they want you to finish.

94

**MAXIM 3:** Every organization is self-perpetuating. Don't ever ask an outfit to justify itself, or you'll be covered with fact, figures, and fancy. The criterion should rather be, 'What will happen if the outfit stops doing what it's doing?' The value of an organization is easier determined this way.

**MAXIM 4:** Try to find out who's doing the work, not who's writing about it, controlling it, or summarizing it.

**MAXIM 5:** Watch out for briefings. They often produce an avalanche. [*Definition:* a high-level snow job of massive and overwhelming proportions.]

**MAXIM 6:** The difficulty of the coordination task often blinds one to the fact that a fully coordinated piece of paper is not supposed to be either the major or the final product of the organization. But it often turns out that way.

**MAXIM 7:** Most organizations can't hold more than one idea at a time... Thus complementary ideas are always regarded as competitive. Further, like a quantized pendulum, an organization can jump from one extreme to the other, without ever going through the middle.

**MAXIM 8:** Try to find the real tense of the report you are reading: Was it done, Is it being done, or Is it something to be done? Reports are now written in four tenses: past tense, present tense, future tense, and pretense. Watch for novel uses of CONGRAM (CONtractor, GRAMmar). Defined by the *imperfect past*, the *insufficient present*, and the *absolutely perfect future*.

# On the optimum size for an establishment

*H W H O Petard*

RBP of Maurice
Vincent Wilkes

## 1. *Introduction*

It is commonplace that the amount of work done by an Establishment of Civil Servants is not directly proportional to its size; indeed, cases may readily be quoted in which an increase in staff has impeded rather than furthered the objects for which an establishment exists.

In the present paper I shall examine the problem analytically, and derive formulae by means of which the optimum size for an establishment may be determined. I shall make use for this purpose of the *Kinetic Theory of Civil Servant Swarms*, which has been developed by Ponticelli. As this theory is not as well known as it deserves, I propose in the next section to give a brief summary of the most important results.

## 2. *The Kinetic Theory of Civil Servant Swarms*

The fundamental concept in this theory is 'pressure', which may be defined as the force that the establishment as a whole can bring to bear in order to achieve its objectives, for example, to overcome the obstruction of other establishments or departments, or itself to obstruct some design originating from without. At first sight it might be thought that a Civil Service Establishment, by virtue of the random motions of its members, would exert a pressure of precisely zero; a moment's consideration of the analogous case of a gas consisting of molecules moving with random velocities will, however, show that this is not the case, and that in fact a finite pressure is always exerted. This pressure is, of course, less than the pressure of a hypothetical pseudo-establishment in which the energies of the staff are all oriented in the same direction.

I will not here enunciate all the results of the Kinetic Theory, for which reference should be made to the original sources; two results, however, are of particular importance.

> (*a*) In order to do work, an establishment must continually expand.

(b) The larger the establishment becomes, the less is the pressure it exerts.

### 3. *The Optimum Size for an Establishment*

We shall suppose that the establishment consists of $n$ members; this number is only to include those members of the staff in positions of responsibility, as it is not my intention in this paper to examine in detail the relations existing, for example, between an officer and his secretary. Let a fraction $k$ of each officer's time (office hours only being considered) be devoted to the work of the establishment. (We shall not consider what use is made of the remaining fraction $(1 - k)$ as it is felt that this could be better dealt with elsewhere, and by other methods.) And let a fraction $m$ then be devoted to work other than internal liaison with other members of the establishment. We will suppose for simplicity that the quantities $k$ and $m$ are the same for all members. Normally, $k^2$ and higher powers may be neglected. In what follows it will be convenient to take the working day as the unit of time.

Since each officer has to liaise with each of his $(n - 1)$ colleagues, we have, assuming he spends the same time with each,

$$k - m = a(n - 1)$$

where $a$ is a constant. The total time spent on liaison by all the officers taken together is then

$$an(n - 1)$$

while the total time spent on useful work is

$$nm.$$

It will thus be seen that as $n$ increases, the time spent on liaison increases much more rapidly than the time spent on useful work. If we assume that the size of the establishment is optimum when these two quantities are equal, we

97

have

$$an(n - 1) = nm$$
$$\therefore n = 1 + m/a.\dagger$$

Since exactly half the time $k$ is spent on useful work we must have

$$m = k/2$$

so that the above equation for $n$ becomes

$$n = 1 + k/(2a).$$

## 4. A Numerical Example

To illustrate the foregoing results in a practical case, let us suppose that daily liaison between officers is limited to the writing of one short minute, or the making of one short telephone call, the total duration being on the average 3 minutes. If the length of the working day is 5 hours, we have

$$a = 3/(5 \times 60) = 0.01.$$

If we now take for $k$ the value 1/2, we get for $n$

$$1 + (1/2)/(2 \times 0.01)$$

that is 26.

Thus the optimum size for an establishment under these conditions is 26.

---

†We reject, regretfully, the solution $n = 0$.

# Research

Every great scientific truth goes through three stages. First, people say it conflicts with the Bible. Next, they say it has been discovered before. Lastly, they say they always believed it.

*Louis Agassiz*

# Mathematical analysis of the twelve days of Christmas

Contributed by
John N Shive.

| Item | No of items given each time | No of days given | Total No of each item |
|------|------|------|------|
| Partridge in a pear tree† | 1 | 12 | 12 |
| Turtle doves | 2 | 11 | 22 |
| French Hens | 3 | 10 | 30 |
| Calling birds | 4 | 9 | 36 |
| Golden rings | 5 | 8 | 40 |
| Geese-a-laying | 6 | 7 | 42 |
| Swans-a-swimming | 7 | 6 | 42 |
| Maids-a-milking | 8 | 5 | 40 |
| Ladies waiting | 9 | 4 | 36 |
| Lords-a-leaping | 10 | 3 | 30 |
| Pipers piping | 11 | 2 | 22 |
| Drummers drumming | 12 | 1 | 12 |

The data in the 4th column above is shown in the plot. The curve‡ is an inverted parabola given by the equation:

$$T = 42\tfrac{1}{4} - (N - 6\tfrac{1}{2})^2. \tag{1.1}$$

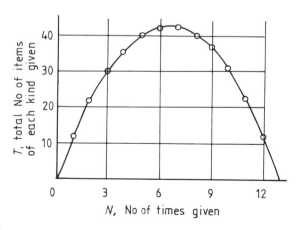

The curve has roots at the 0*th* and 13*th* days. On these days nothing was given. The curve is real, finite and single-valued only in the region

$$0 < N < 13.$$

To determine the time at which the function has zero slope, differentiate Eq. (1.1) and set result equal to zero.

$$dT/dN = -2(N - 6\tfrac{1}{2}) = 0 \qquad (1.2)$$

from which $N = 6\tfrac{1}{2}$, the axis of the parabola.

To find the total of all items given, integrate under the curve of Eq. (1.1)

$$N_{tot} = \int_0^{13} T \, dN = \int_0^{13} 42\tfrac{1}{4} \, dN - \int_0^{13} (N - 6\tfrac{1}{2})^2 \, dN = 364\tfrac{1}{4}.$$

$$(1.3)$$

By actual count:

$$N_{tot} = 364 \ \S.$$

†One partridge in one pear tree is counted as one item given.
‡Projections of the curve into regions of negative giving (i.e. taking away) are not allowed. Only positive giving is considered.
§The discrepancy of one quarter of an item is accounted for as the difference between the integral and the discrete series.

---

## An anthology of Einsteinian limericks

**On the ultimate speed**

Contributed by
A Anderson.

*'If I could move faster than light'*
*Mused Einstein, when a lad so bright;*
*'I could set off one day,*
*In a relative way,*
*and return on the previous night!'*

### On gravitation

*To Newton, and to most of the race,*
*Gravitation is just one special case*
*Of forces which obey*
*F = ma;*
*But to Einstein it results from curved space.*

### On relative motion

*When Einstein was travelling to lecture in Spain,*
*He questioned a conductor time and again:*
*'It may be a while,'*
*He asked with a smile,*
*'But when does Madrid reach this train?'*

### On electron – positron annihilation

*A scientist named Lee wrote a note on*
*A way to change mass into photon.*
*He showed Einstein his data,*
*But he made 'light of the matter'*
*And said it was nothing to gloat on!*

---

# Quoted and attributed in *The Observatory*

RBP of the
Editors of *The Observatory*.

No 1031 August 1979.

*Beyond the Mohole*
Apart from completing the first circumnavigation of the world along its polar axis, the main aims of the expedition are...—*Daily Telegraph*, February 17, 1979

\* \* \* \* \* \*

*Things that Go Bump in the Night*
...the planets have paths which are more or less in the same place.—P Moore, *The Comets, Visitors from Space* 1973 p 22

******

*Theoreticians Might Not Be so Self-effacing*
Although the personalities of individual observers are subject to a considerable uncertainty ... the mean personality—by definition—is zero.—*A. & An.* **61** 225, 1977

******

*Oubliette?*
The walls and ceiling of the annexed premises in the former Chemical Society building have now been put in good order. Toilet facilities have been provided. We understand that certain library material is to be taken to these rooms to ease the present congestion of shelf and storage... —E L G Bowell and C Jordan *Q. J.* **11** 161, 1970

******

*Hell fire?*
Temperature in the centre of the planet reaches probably 300,000 degrees on the Calvin scale... —*The Guardian* June 10, 1977

******

*We had supposed they had been removed by Caesarean!*
Excess days accumulated in the Julian calendar were to be exorcised... —Sir Fred Hoyle, *Astronomy and Cosmology* W H Freeman, 1975 p 96

******

*When UPS Goes Down*
The uninterruptible power supply (UPS) used to provide power to the computer data system failed... —*KPNO Quarterly Bulletin* July/Dec 1977 p 31

*Sources Close to the Divine?*
Simmons finds the month of the birth of Christ by relating the lives of Christ and John the Baptist (personal communication). —*Nature* **268** 566, 1977

\* \* \* \* \* \*

*Twinkle, twinkle little star*
*How I wonder what you are,*
*Dithering above so high*
*Like a dancer in the sky.*
*For your disc so bright and Airy*
*Is so very apt to vary,*
*That it's hard to say just when*
*You cross the true meridian.*

\* \* \* \* \* \*

## The Irishman's astronomy

*Long life to the Moon, for a dear noble cratur,*
*Which serves us for lamplight all night in the dark.*
*While the Sun only shines in the day, which by natur*
*Wants no light at all, as ye all may remark.*

---

# The expanding unicurse

Blanche Descartes *Proof Techniques in Graph Theory* (New York: Academic Press) 1909 p 25. Copyright © 1909 by Academic Press, Inc. Reproduced by permission.

*Some citizens of Königsberg*
*Were walking on the strand*
*Beside the river Pregel*
*With its seven bridges spanned.*

*'O Euler, come and walk with us,'*
*Those burghers did beseech.*
*'We'll roam the seven bridges o'er,*
*And pass but once by each.'*

*'It can't be done,' thus Euler cried.*
*'Here come the Q. E. D.*
*Your islands are but vertices*
*And four have odd degree.'*

104

*From Königsberg to König's book*
*So runs the graphic tale*
*And still it grows more colorful*
*in Michigan and Yale.*

---

## Human genetics

[Professor F Clarke Fraser, MD, presented this prescient verse at an Allan Award banquet of the American Society of Human Genetics in 1963.]

It would be a nice tradition for each President to make some sort of bequest to forward the work of the Society. The following contribution is therefore presented in readiness for the day when it will be possible to change the genetic constitution of man by DNA transformation. If the Society is on its toes it will patent the process, corner the market on human DNA, and conduct a vigorous advertising campaign, which will of course require a jingle for the TV commercial. This one is sung to the tune of 'Smiles.'

*There are genes that make you happy*
*There are genes that make you blue[1]*
*There are genes that tell you who's your father[2]*
*And how you'll rate on your I.Q.*
*There are genes that make your blood clot quickly*
*And genes that tell how much you'll weigh[3]*
*But if you don't like the genes you're born with*
*TRY A.S.H.G. D.N.A.[4]*

[1]Congenital methemoglobinemia, for instance.
[2]Poetic license; actually, of course, they can usually only tell who's *not* your father.
[3]If you don't make a pig of yourself.
[4]Copyright pending.

# Lines inspired by a lecture on extra-terrestrial life

J.D.G.M. in *The Observatory* 65: No 815, August 1943 p 87–88. RBP of the Editors of *The Observatory*.

*It seems improbable that one*
*Could live in comfort on the Sun.*
*The temperatures encountered there*
*Would baffle any Frigidaire*
*And its inhabitants are cursed*
*With pangs of so intense a thirst*
*That every thought and word and wish*
*Impels them to imbibe like Fish.*
*Research has all too clearly shown*
*The squalor of their moral tone.*
*One cannot keep to social rules*
*When shaky in the molecules*
*And nuptual bliss must needs take wings*
*When even atoms lose their rings.*
*How limited, in fact, the scope*
*Within a gaseous envelope!*
*I knew a man who, undeterred*
*By all the things that he had heard,*
*Deserted an attractive wife*
*To plunge into the Solar Life.*
*He had been told—one hopes, in fun—*
*That there are spots upon the Sun;*
*Which proved, he said, that at the worst*
*One simply couldn't die of thirst*
*And at the best might daily dine*
*On draughts of Infra-ruby wine.*
*He waited till the time had come*
*When spots are at the maximum,*
*And, while his wife was out at church,*
*Slipped off—a martyr to Research.*

\*\*\*\*\*\*

[From a Chinese astronomer]

*A moon, a nightingale and Her—*
*How different on Jupiter!*
*Nine moons across his heaven dance—*
*What invitation to Romance,*

106

To nightingales and their sweet plaint
(If there were any, — which there ain't)
To maids aglow with looks aslant
(If there were any, — which there aren't).
Nine moons, No nightingales, No Her—
How terrible on Jupiter!

******

Some time ago my late Papa
Acquired a spiral nebula.
He bought it with a guarantee
Of content and stability.
What was his undisguised chagrin
To find his purchase on the spin,
Receding from his call or beck
At several million miles per sec.,
And not, according to his friends,
A likely source of dividends.
Justly incensed at such a tort
He hauled the vendor into court,
Taking his stand on Section 3
Of Bailey 'Sale of Nebulae.'
Contra was cited Volume 4
of Eggleston's 'Galactic Law'
That most instructive little tome
That lies uncut in every home.
'Cease' said the sage 'your quarrel base:
Lift up your eyes to Outer Space.
See where the nubulae like buns,
Ancurranted with infant suns,
Shimmer in incandescent spray
Millions of miles and years away.
Think that, provided you will wait,
Your nebula is Real Estate,
Sure to provide you wealth and bliss
Beyond the dreams of avarice.
Watch as the rolling aeons pass
New worlds emerging from the gas:
Watch as the brightness slowly clots

To eligible building lots.
What matters a depleted purse
To owners of a Universe?'
My father lost the case and died:
I watch my nebula with pride
But yearly with decreasing hope
I buy a larger telescope.

---

# Note on the age of the earth

## (the earth is 8556249847569.7 years old.)

Emery Meschter
The Gamma
Alpha Record **26**
No 3, May 1936
p 77–78

The question of the age of the earth has long been an intriguing one for scientists. Calculations based on the rate of decay of radioactive substances and also upon the composition of sea-water have given us a rough idea of the time which has elapsed since the earth took on something like its present form. It appears, however, that certain rather simple considerations which have thus far been overlooked can furnish us with additional estimates. It is the purpose of this paper to show how several determinations of the age of the earth may be made, employing only the most elementary mathematics and physical theory.

According to data compiled by the International Institute of Meteorology at Omsk, about 37% of the earth's surface is covered by clouds, which may extend to a maximum height of ten miles. We must recall that the air must be saturated with water vapor before condensation occurs and clouds are formed. At 20°C saturated air contains 0.4875 grams of water vapor per cubic foot. The total weight of water required to form these clouds, therefore, is $5.336 \times 10^{19}$ grams, remembering that the earth's surface comprises $3.502 \times 10^8$ square miles. But in order to cause water to evaporate we must supply heat to the extent of 582.2 calories per gram (at 20° C), and in the case of the clouds which we are considering this heat must have come from the sun. Experiments show that

$1.3 \times 10^6$ ergs of radiant energy fall each second upon each square centimeter of the earth's surface upon which the sun shines. The total heat energy received by the earth each second may thus be easily calculated by means of arithmetic, yielding a result of $1.247 \times 10^{16}$ calories. Combining these results, we find that it would take $2.495 \times 10^6$ seconds to vaporize $5.336 \times 10^{19}$ grams of water requiring 582.2 calories per gram, by supplying heat at the rate of $1.247 \times 10^{16}$ calories per second. This is equivalent to 28 days, 21 hours, 7 minutes, and 30 seconds, a result somewhat lower than estimates made by previous investigators. However, since all methods involve a great deal of uncertainty, it must be given due weight in determining our final judgment as to the age of the earth.

It is a well-established fact that the earth is not exactly at zero potential, although it is usually assumed to be for the sake of convenience. Measurements show that its potential is actually 50.672 volts negative; since the capacitance of the earth is 710 microfarads, a charge of $3.5997 \times 10^{-2}$ coulombs would produce this result. The obvious source of charge is lightning strokes, one of which occurs, on the average, every sixty-seven seconds. Although these flashes consist of currents as large as $10^4$ amperes their time of duration is extremely short—about $10^{-6}$ seconds. Hence, with each discharge $10^{-2}$ coulombs are transferred to the earth, and the time required to charge the earth to its present potential is 4.99 minutes. Again we have obtained a result which may seem rather low, but rather than be discouraged let us turn our attention to a somewhat more reliable method.

Sixty-eight per cent of the surface of the earth is covered by the several seas to an average depth of 12,842 feet; were this distributed uniformly over the earth, we would have an ocean 8,713 feet deep. The source of all this liquid must obviously be rain; turning once again to the figures of the Institute of Meteorology, we find the annual rainfall averaged over all the earth is 31.22 inches. The time required, therefore, to fill the ocean to its present level is easily found. Upon completing this calculation we discover that the age of the earth is 1,409 years. Greater accuracy than one year cannot be realized by this method, due to the uncertainties of the rainfall data. It is interesting to note at this point that our latest result is of the same order of magnitude as the estimates of the oldest writers on the subject, who give the age of the earth as about 6,000 years.

Perhaps the most accurate and fundamentally correct method of all is to consider the actual source of the 'stuff' of which the earth is composed. Although we receive great quantities of radiation by way of inter-stellar space, comparatively little matter comes to us in this way: an occasional meteorite and a relatively small number of electrons. What, then, is the source of the earth's matter? Casting about for light on this question, we can discover but one other possible source, which is even to this day active in building up the earth. I refer, of course, to volcanoes, a great number of which are still active. These pour out $6.7829 \times 10^9$ tons of lava each year, so that the mass of the earth is annually increased by this amount. The total mass of the earth is known to be $5.799 \times 10^{21}$ tons; hence it must have required 8556249847569.7 years to build the earth up to its present mass. It will be seen at once that this is by far the most magnificent result which has been obtained by any investigator up to the present time.

Our final conclusion as to the true age of the earth must be an average of the several results just obtained. Combining the four in this way, we arrive at the figure of 2134062461892.425 years which have elapsed since the beginning of the earth as we know it today.

# Said a mathematician (age 7)

D C Perfect in
*The Observatory*
73: 216, 1953.
RBP of the
Editors of *The
Observatory*.

*Said a mathematician (age 7)*
*'Shall I ever get into Heaven*
*If I cannot tell why*
$hc/\pi$
*$2e^2$ must be 137?'*

*Said Father, 'I very much doubt*
*If you know what you're talking about,*
*But since Heaven, my boy,*
*Is a place to enjoy,*
*The Righteous will see you're kicked out.'*

*'Oh Mathykins, what an illusion!*
*You forget that the Law of Exclusion*
*Keeps, as Pauli observed,*
*Heaven's places preserved*
*From the victims of quantum confusion.'*

*His parents replied with some gravity*
*'You'll certainly be hung for depravity*
*if you say that pro tem*
$c^3 t/M$
*Is $\gamma$ inside of a cavity.'*

---

## Raman's rogues

Contributed by Subramanian Raman.

I showed these pictures (in sequence) at the International Conference on Reactions between Complex Nuclei, Vanderbilt University, Nashville, 1974, and elicited a good laugh.

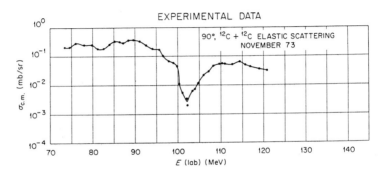

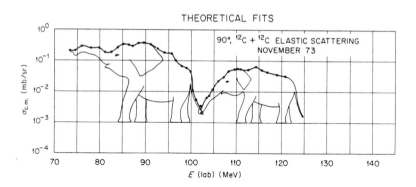

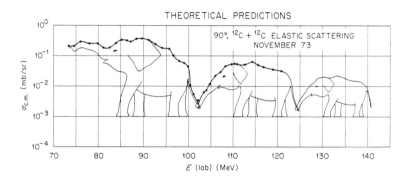

---

# 1 + 1 = 2 expounded

*Anon*

Contributed by
Lawrence T Scott

Every new scientist must learn that it is never in good taste to designate the sum of two quantities in the form

$$1 + 1 = 2. \tag{1}$$

Anyone who has made a study of advanced mathematics is aware that $1 = \ln e$ and $1 = \sin^2 x + \cos^2 x$. Further,

$$2 = \sum_{n=0}^{\infty} \frac{1}{2^n}$$

therefore Eq. (1) can be expressed more scientifically as

$$\ln e + (\sin^2 x + \cos^2 x) = \sum_{n=0}^{\infty} \frac{1}{2^n}. \tag{2}$$

This may be further simplified by use of the relations

$$1 = \cosh y \sqrt{1 - \tanh^2 y}$$

and

$$e = \lim_{z \to \infty} (1 + 1/z)^z.$$

Equation (2) may therefore be rewritten

$$\ln\left[\lim_{z \to \infty}(1 + 1/z)^z\right] + (\sin^2 x + \cos^2 x)$$

$$\tag{3}$$

$$= \sum_{n=0}^{\infty} \frac{\cosh y \sqrt{1 - \tanh^2 y}}{2^n}.$$

At this point it should be obvious that Eq. (3) is much clearer and more easily understood than Eq. (1). Other methods of a similar nature could be used to further expound Eq. (3); these are easily discovered once the reader grasps the underlying principles.

---

# The population density of monsters in Loch Ness

R W Sheldon and S R Kerr 1972 Limnology and Oceanography 17 796–8. RBP of the American Society of Limnology and Oceanography.

It is well known that there are monsters in Loch Ness. Their most characteristic features are that they are rarely seen and never caught, but there are records of sightings extending back many centuries. The fact that they are rarely seen suggests that the population is small. It is known from direct observation that the animals themselves are large and it follows from this that the population *must* be small. It can be demonstrated quite easily from trophic – dynamic considerations that many large animals could not exist in Loch Ness; but a few could. It has been suggested from time to time that as the monsters are never caught it must therefore follow that they do not exist. This is both irresponsible and illogical.

Many accounts have been written of Loch Ness and its monsters (e.g. Holiday 1968) but very few quantitative observations have been made. We know nothing of their distribution. The population structure of the monster community is also unknown to us. As they are rarely seen and never caught (characteristic features) it is particularly difficult to study their population dynamics. However, it is our purpose to show that it is possible to estimate the number of monsters that can exist in Loch Ness.

The production rate of oceanic organisms is size dependent, but in ecologically stable areas the standing stock is constant at all sizes (Sheldon *et al* 1972). It is not unreasonable to assume that similar relationships exist in large bodies of freshwater. If this is so then the standing stock of monsters taken over logarithmic size intervals should be similar to that of other organisms (e.g. fish or plankton).

We have not been able to find any information on the standing stocks of Loch Ness, but an estimate of the fish stock can be made if the probable yield is known. A deep oligotrophic lake such as Loch Ness should give an annual yield of rather less than 1 kg ha$^{-1}$ yr$^{-1}$. This estimate can be refined by calculations based on Ryder's (1964, 1965) morphoedaphic index (total dissolved solids/mean depth). Again, we could not find data from Loch Ness and have used a value for total dissolved solids for the northern part of Loch Lomond (Darling and Boyd 1969). The estimate of mean depth was taken from Hutchinson (1957). By using this information in Ryder's (1964) equation we calculate that Loch Ness should give an average fish yield of 0.55 kg ha$^{-1}$ yr$^{-1}$. The ratio of biomass to production of a fish producing system will range from about 1 to 5, so that the standing stock of fish in Loch Ness should lie in the range from 0.55 to 2.75 kg ha$^{-1}$. The concentration of monsters should be similar.

The area of Loch Ness is about 5,700 ha. The total mass of monsters in the loch is therefore in the range 3,135 to 15,675 kg. In figure 1 we show the number of monsters the loch could support relative to individual size. The minimum average size is taken arbitrarily as 100 kg: anything smaller is not suitably monstrous. The number of monsters in the loch could vary from 1 to 156 depending on the standing stock and average size. The largest number would occur in the situation where high standing stock and small average monster size coincide; however, we believe that such a situation is unlikely. The smallest number must be more than two if the species is to be maintained. Monsters have been seen in the loch for hundreds of years so that there must be a breeding population. The alternative possibility, a single monster of great age, is unlikely, and *inter alia* is not in keeping with the wide range of size estimates reported in the literature. A viable population could be quite small but probably would not be less than 10. This constraint is indicated by the vertical line in figure 1. All the combinations of individual monster weight and population shown by figure 1 are theoretically possible, but we would only consider those to the right of the vertical line to be realistic.

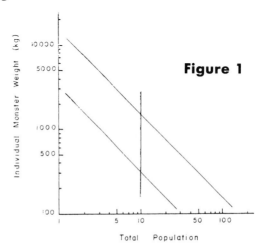

**Figure 1**

115

We will now attempt to show that some of the individual monster weight and population combinations are more probable than others. Much of our reasoning is based on observational evidence.

The trophic position of the monsters is probably that of terminal predators feeding on fish (Holiday 1968). The growth efficiency of many aquatic predators is around 10%. If the monsters are similarly efficient and if a major part of the fish production is used by them, then their production must be of the order of 300 kg yr$^{-1}$ or more. The average number of deaths per year is determined in a stable population by the ratio of production to mean size. On this basis monsters weighing 100 kg would have to die at a minimum rate of about 3 per year. Larger monsters would die less frequently.

Two lines of evidence support the view that monsters do not die frequently and must therefore be large. Firstly, corpses are never found. Secondly, a relatively large number of juveniles must exist if adult mortality is high, but although small monsters have been seen from time to time they are not common. It seems therefore that Loch Ness must contain a small number of large monsters. These could weigh as much as 1,500 kg with a population of 10–20 individuals. A 1,500 kg monster could be about 8 m long, a size that agrees well with observational data.

We are aware that in these calculations we have not taken migratory fish into consideration. These will increase the effective standing stock of the loch and this could result in there being either more or larger monsters than we have shown. However, Sheldon *et al* (1972) suggest that standing stocks are not absolutely constant. There is probably some decrease at the higher trophic levels which could result in there being either fewer or smaller monsters than we have shown. These two factors are antipathetic, and although we do not know the relative magnitudes, they are both likely to be of the order of a factor of two. They will tend to cancel each other out and it is not improbable therefore that the population density that we have described for the monsters in Loch Ness is near to the true value.

It is not unknown for sightings of monsters, both in Loch Ness and elsewhere, to go unrecorded (Heuvelmans 1968, Holiday 1968). Fear of ridicule is the main reason why many observers do not make their observations known to science. But it is the skeptics who are at fault. Monster observers should be encouraged. The occurrence of monsters is quite reasonable and by no means fantastic.

We would like to thank Kate Kranck for drawing our attention to this problem, because until she mentioned it we were unaware that monsters were a problem.

## Watchwords

RBP of Clifford E Swartz.

*The Physics Teacher* has, at Christmas time, offered its readers watchwords 'suitable for framing.' Three are reproduced here for the December issues for 1969 and 1971.

Regarding the first, which might seem outrageous, the Editor, Clifford Swartz, remarked: 'Remember that physics is not necessarily the science of precision; it is preeminently the science of common sense. To do a thing well enough for the purpose at hand may well require a lifetime of skilled and devoted work.'

| | | | | | |
|---|---|---|---|---|---|
| 0.0003456 | $\sigma$ | 1,843,625 | 4,000,000,001 | $5.64 \times 10^{-8}$ | $\sigma$ |

1 part in $10^4$

$\sqrt{\dfrac{\sum n_i (t_i - \bar{t})}{N}}$

$x^2$

±89

$\bar{t}$

### If a thing is worth doing, it's worth doing well

*enough for the purpose at hand— and it is surely silly and probably wrong to do it any better.*

$\bar{t}$

±0.5%

$x^2$

±0.08

$5.000 \times 10^3$

5000

84 ± 6

$\bar{t}$

| | | | | | |
|---|---|---|---|---|---|
| $\bar{t}$ | 83.4567 ± 15.1 | $\sigma$ | 1 part in $10^4$ | $\sqrt{\dfrac{\sum n_i (t_i - \bar{t})}{N}}$ | ±15% |

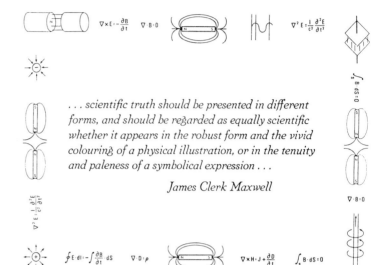

> ... scientific truth should be presented in different forms, and should be regarded as equally scientific whether it appears in the robust form and the vivid colouring of a physical illustration, or in the tenuity and paleness of a symbolical expression ...
>
> James Clerk Maxwell

$$\mathcal{A}\text{nd God said:}$$

$$\nabla \cdot \mathbf{E} = \frac{\rho}{\varepsilon_0}$$

$$\nabla \cdot \mathbf{B} = 0$$

$$\nabla \times \mathbf{E} = -\frac{\partial B}{\partial t}$$

$$c^2 \nabla \times \mathbf{B} = \frac{j}{\varepsilon_0} + \frac{\partial E}{\partial t}$$

*—and there was light*

# Are you troubled by irregularity?

*Denis Weaire*

From the 'Lateral Thoughts' section of *Physics World*, June 1989 (Bristol: Institute of Physics Publishing Ltd).

In the days of my irresponsible youth, now faded into irresponsible middle age, I first developed a professional interest in the theory of amorphous solids, poor cousins of their crystalline counterparts. To arrest the attention of seminar audiences in the US, I used to begin with a portentously solemn introduction: 'Friends, are you troubled by irregularity?' This was taken from a laxative commercial, familiar at that time. Irregularity of all sorts was indeed troubling the solid state theorists. More traditional types could not accept that the medical world is not organised entirely along the lines of the Grenadier Guards.

How different now. It has become high fashion, positively *de rigueur*, to analyse disorder. Many such as myself have made a comfortable living out of pursuing it in every form: temporal, spatial, compositional, topological...

Unfortunately, disorder also rules in certain aspects of my private life. I am particularly afflicted by the phenomena manifested by my extensive collection of socks. Recognisable pairs among the random ensemble are quite a rarity. The Pauli Principle applies: no two identical socks can co-exist in the same place. Even large banks of coincidence counters would register only a desultory click, once a fortnight, in my bedroom.

This understates the problem. Even broadly similar socks prove difficult to find in that environment. There must be a strong repulsive interaction between them. On the other hand, dissimilar socks cohere pretty well. I reckon it must take a force of several newtons to separate a calf-length Hunting Sutherland pattern wool-nylon combo from a brushed cotton striped Jamaican beach sock.

Were I to throw out the offending singles, I know that their twins, no longer experiencing powerful forces, would come back to haunt me. If I buy more socks, entropic effects will soon decrease the coincidence count yet further. So what has theory to offer? As usual, not much. The mean field approximation leads quickly to the pronouncement 'All socks are grey', but this holds only in

119

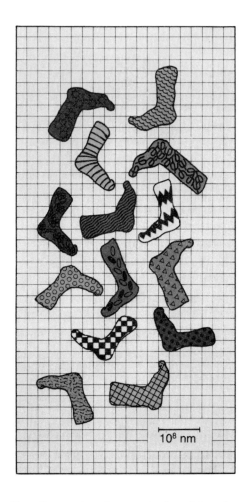

**Figure 1.** Small sample of a typical multi-sock array, in the normal (or Weaire) state, manifesting topological, orientational and compositional disorder.

Eastern Europe and some of the lesser public schools. Sufficiently frayed socks are, they tell me, fractal in character: I find this equally unhelpful.

No, a technological breakthrough is required. Cooper Pair Socks, which stick together, would fit the bill. Or insurmountable quark-like attractions (I could always carry the third sock as a spare). Who knows, somewhere in Utah...

In the meantime, what's to be done? Mostly I rely on sang-froid. I stride off into the dawn with jutting chin and odd socks. My claim that this is the latest thing, dictated by fresh intelligence from the King's Road, is generally derided, at least when the full ensemble is completed by double-breasted pinstripe and college tie.

While waiting for fashion to catch up with me, I am consoled to realise that I do not suffer alone. Why else would 'Sock Brokers' have sprung up at all our major airports? Those 'executive saddlebags' harbour mismatched hosiery, to be surreptitiously replaced before take-off, you may be quite sure. So I may not, after all, require a socks counsellor, or words of comfort from Dr Alex Comfort, in a fully illustrated blockbuster, *Joy of Socks*. He would be too late anyway; the entire field of trendy disorder has just been launched into the public consciousness by James Gleick (*Chaos — Making a New Science*) and Thomas J Peters (*Thriving on Chaos*). The latter is aimed at executives whose filing systems have degenerated into a random state. I bought a copy to wave at people who made adverse comments on the state of my office. If only I could find it...

These subversive books are bound to transform society, so why not come out of the closet and do some spontaneous symmetry breaking? Let your ankles assert their individuality. Together we can build a Random Republic. In such a disordered state, trains will run at random times. This will require little adjustment by British Rail and take all the angst out of travel at a single stroke. Soccer players will wear individual uniforms, making the judgement of offsides, which is the primary interest of the game, more exciting. Universities will award random degrees. (Harassed VCs please note.)

There will be three days of amnesty during which all anal compulsives may hand in their Filofaxes and trouser presses at the nearest police station. Group theorists will be offered early retirement. Get ready for D-day.

# On the electrodynamics of moving fireflies

Contributed by
D T Workman.

**Abstract.** A theory is proposed in which the source of the firefly light is due to the inclusion of wintergreen plants in its diet. An empirical equation describing the light emission process is developed, and several neat conclusions are drawn.

One of the most enduring problems of modern science has been the attempt to explain the source of the light produced by the firefly. In attempts to solve the problem, millions of fireflies have been captured, crushed, and chemically analysed, all to no avail! Most theoretical work has centered around various modifications of the Ising model of pseudo-spin interactions utilizing a non-harmonic cross-term connecting the symmetric and non-symmetric states of the elliptically polarized non-degenerate wave functions describing the firefly's tummy [1–4]. These have been able to predict light emission, but it always comes out the wrong color.

However, as a result of a typically serendipitous event, quite analogous to others in the history of the development of scientific ideas [5], we have been able to propose the actual mechanism which produces the firefly's light. The author happened to observe several of his students in G.S. 152D relaxing during a break in the class. They were in a dark room crunching wintergreen Life-Savers and had observed that a small flash of light is produced with each crunch. They seemed impressed by this fact, but did not realize the cataclysmic implications. However, demonstrating the well-known law that 'chance favors the prepared mind', the author immediately recognized the truly significant aspect of this phenomenon—the light given off was exactly the color of that produced by fireflys!

Immediately, further investigation of this fascinating phenomenon was initiated. It was first confirmed that the light would be produced when crunching was performed by an unbiased person under controlled conditions, i.e., by the author in his bathroom. The investigation was then expanded by constructing an automatic crunching device, hereafter called 'The Cruncher', to enable the investigation to proceed to more quantitative analysis. This was accomplished using computer controlled servomechanisms with solid state diode rectified transducers for measurement of pressure gradients [6]. Using this device, it was ascertained that no light was produced by crunching peppermint or fruit-flavored Life-Savers. Neither was any light produced by Certs with retsyn, Chiclets, gum-balls, jaw-breakers, or Brach's Chocolates. The phenomenon seems limited to wintergreen Life-Savers.

The last piece of the puzzle came to the author while at a performance of Stravinsky's ballet suite *The Firefly* [7]. The secret is the wintergreen! Immediately a supply of wintergreen leaves and berries were procured and tested in 'The Cruncher'. On the first crunch, a tiny flash of light was observed. Surely Heinrich Hertz could not have been more gratified at his first production of a spark from electromagnetic waves than we were at this historic moment! Further research has enabled us to develop an empirical equation describing light production from the wintergreen plant (equation 1—hereafter labeled CE, the 'Crunch Equation')

$$I = Cp^{3/2}\, e^{3n/kT}$$

where $I$ = light intensity in $\mu W/m^2$, $p$ = pressure in dynes/cm$^2$, $n$ = number of molecular collisions per second—the crunch frequency (CF), $k$ = the well-known Boltzmann constant, and $T$ = the temperature in degrees Kelvin. $C$ is a constant weakly dependent upon the species of wintergreen used and of the order of unity. The color of light emitted is strongly dependent on the part of the plant used. Leaves give off light rich in green spectral components, while the berries give out a reddish-yellow color light. Both of these components must be present in the chartreuse light of the firefly.

Since the Crunch Equation (CE) shows light production to be strongly dependent upon pressure and crunch frequency (CF) the actual mechanism of light production in the firefly seems intuitively obvious. The firefly ingests parts of the wintergreen plant which are then transmitted through its gullet [8] into a strongly muscular masticator (quite analogous to a chicken's gizzard) located in the firefly's lower abdomen. Here the wintergreen leaves and berries are 'crunched up' giving off in the process the light that has baffled us for so long. An order of magnitude calculation indicated that light intensities would not be intense enough if wintergreen plants alone were present during mastication, leading to the additional proposal that fireflies must ingest rocks to intensify the mastication (and light producing) process, a phenomenon also observed in chickens and proposed for dinosaurs [9].

Research into our proposal is continuing and several lines of inquiry would seen to be most profitable:

(1) dissection of a firefly in an attempt to find the masticator should be initiated at once (this is a procedure ill-befitting a physicist, so we will leave it to our biologist friends);

(2) a firefly observed ingesting rocks would be positive evidence for our hypothesis.

(3) our theoretical analysis has been thwarted by the inability to diagonalize our wintergreen wave function (mathematicians—HELP!);

(4) if wintergreen extract can be procured in sufficient quantity, it could be used in conjunction with a simple Squeezer for light production in the home, possibly alleviating the current energy shortage;

(5) since the light given off by crunching a wintergreen Life-Saver is the *same* color as that produced by the firefly, it seems obvious that Life-Savers Inc is using *leaves* in their product instead of just *berries*. We suggest an immediate investigation by the Federal Bureau of Consumer Affairs.

*References*

[1] I M Wun and U R Tu, 'The Firefly—Chemistry's Failure,' *Journal of Tibetan Chemistry* vol 74, no 6, November 23, 1975, p 761

[2] Ha-Ha Haha, 'Anomalous Kerr Effect and Firefly Light,' *Royal Physical Society of Pago Pago* vol 74, no 6, November 24, 1975, p 762

[3] Dyne O Mite, Firefly Light and Zero-order Game Theory,' *Journal of Obscure Mathematics* vol 74, no 6, November 25, 1975, p 763

[4] I M Wong, 'Group Theoretical Analysis of Firefly Light,' *Journal of Irreproducible Results* vol 74, no 6, November 26, 1975, p 764

[5] Nick D Greek and L Luck, 'Pure Blind Chance and Scientific Discovery,' *Journal of the History of Science in America in the Post-War Years* Part D, Mississippi Section, vol 74, no 6, November 27, 1975, p 765

[6] Blueprints available from author for a small fee

[7] Performance 74, part 6, November 28, 1975

[8] *All About Crawly Things—A Toddler's Guide* (Any town, U.S.A.: Wee-Ones Press), 74th Printing, part 6, p 766

[9] James A Michener, *Centennial* (New York: Fawcett Books) November 1975, p 74

# Interdisciplinary Strivings

People will tell you that science, philosophy, and religion have nowadays all come together. So they have in a sense...they have come together as three people may come together at a funeral. The funeral is that of Dead Certainty.

*Stephen Leacock*

# How to write an article to impress your peers

Marleen Boudreau Flory *The Chronicle of Higher Education* January 26, 1983. Abridged. Copyright 1978, The Chronicle of Higher Education. Reprinted with Permission.

Recent Ph.D.'s who have just begun to write articles for publication may be ignorant of secrets well known to experienced and established scholars. The following set of guidelines is meant to help these young ignoramuses, who may have the outdated notion that scholarly articles ought to have interesting theses, be clearly written, and present new ideas. Young scholars who foolishly persist in such thinking had better start filling out applications for law school.

• *The title.* Try to hide the subject of your article from the reader. After all, you don't even know him; why should you help him? Use obscure language. Make the title as long as possible. Abstract nouns with qualifying adjectives are particularly suitable: 'Semantic Differentiation and Symbolic Symbiosis in the Tragedies of Shakespeare' (for an article about the conjunction 'and').

Hyphens tend to impress: 'Pre-pubescent and Post-adolescent Emotion-centralization'—and, of course, they have the virtue of confusing your reader further.

A title might also include the word 're-examination,' or for those interested in a slightly dry and witty note, 'again'. Articles of this sort should always end: 'And so we see that Lord's theory is probably correct after all.'

• *The footnotes.* Footnotes are far more important than the text of the article. All your readers will read the footnotes, but only a few the article itself.

The first footnote is especially useful for fawning on the famous, most of whom you need to know. Thank everyone you have ever met or heard of who writes on your subject—prehistoric faucet spigots from the upper Nile Delta.

You can praise those who could potentially help you in your career for 'the inspiration of a monumental (or seminal or ground-breaking or epoch-making)' article or book on Nilotic spigots. Thank the old graduate-school professor who gave you 'the courage to pursue the complex subject of faucet spigots in the first place'. Never

128

# Ten mathematical equations that changed the face of the earth

Plate 1

Equations, those statements of relationships among variable factors expressed in the terse shorthand of mathematics, are an indispensable characteristic of physics. In 1971, Nicaragua boldly selected 'the ten mathematical equations that changed the face of the earth' as the theme for a set of stamps on the history of science. The stamps were designed by artists at Thomas de La Rue & Co., British security printers, and were produced by lithography.

Each stamp presents a single equation and suggests pictorially a consequence or application. On the back of each stamp, in Spanish, is a historical reference to the equation and its role in 'changing the face of the earth'. Professor Nicholas M Brentin who made the translations given below says that he suspects, from the somewhat obsolete accentuation and chaotic grammar, that the Spanish text on the stamps is a version of one originally in English.

(Hombre Primitivo)

Elementalmente como es, esta ecuación tuvo consecuencias inmensas para el hombre primitivo porque forma la base de contar. Sin el entendimiento de numeros la gente solamente podría traficar en terminos más rudimentarios; no tenían una tarja exacta del número de ovejas ó vacas que poseían ni cuantos hombres había en su tribu. El descubrimiento de contar condujo directamente al desarrollo rápido del comercio y más tarde a la importante ciencia de medidas.

NICARAGUA   $1+1=2$   CORREO 10 CENTAVOS   DE LA RUE

LAS 10 FORMULAS MATEMATICAS QUE CAMBIARON LA FAZ DE LA TIERRA

**PRIMITIVE MAN** $1 + 1 = 2$

Elementary as it is, this equation had immense consequences for primitive man because it forms the basis of counting. Without the understanding of numbers people would only be able to do business in the most rudimentary terms; they would not have an exact tally of the number of sheep or cows they owned nor of how many men there were in their tribe. The discovery of counting led directly to the rapid development of commerce and later to the important science of measurement.

Plate 2

## PYTHAGORAS 570–479/6 B.C.                    $A^2 + B^2 = C^2$

The most widely used theorem in geometry is undoubtedly that of Pythagoras which refers to the lengths of three sides $a$, $b$, and $c$ of a right angle triangle. It provided for the first time a means of computing lengths by indirect means, allowing man to do surveying and make maps. The ancient Greeks used it to measure the distance of ships at sea, the heights of buildings and other things. Today scientists and mathematicians constantly use it to develop all kinds of theories.

## ARCHIMEDES 287–212 B.C.                    $F_1 x_1 = F_2 x_2$

Archimedes said, 'Give me a place to stand and I will move the world.' The simple equation of the lever is the basis of all engineering, whether it be merely with a crowbar or with the most advanced gearing or crane. Every nut and bolt embodies the principle. The brakes of our cars, scales, door knobs, and the majority of tools are varieties of the lever.

## JOHN NAPIER 1550–1617                    $e^{\ln N} = N$

With the invention of logarithms, Napier gave the world a powerful shorthand for arithmetic. It permitted men to do multiplication or division by simply adding or subtracting the logarithms of numbers so they could carry out rapidly these and more complicated operations containing many numbers. The impact of logarithms in such fields as astronomy and navigation was enormous and is comparable with that of the computer of today's science.

Plate 3

## SIR ISAAC NEWTON 1642 – 1727 $\qquad f = Gm_1m_2/r^2$

Before Newton's time people had very scant idea of what force kept the planets in their orbits around the sun, or the moon around the earth — or even what kept men from flying off the surface of the earth into space. Newton demonstrated that all bodies are attracted to one another by the force of gravity. The equation shows that the force depends on the masses of the bodies. Gravitational force between two small objects is not noticed because it is very weak.

## JAMES CLERK MAXWELL 1831 – 1879 $\qquad \nabla^2E = \dfrac{ku}{c^2}\dfrac{\partial^2E}{\partial t^2}$

A century ago this Scottish physicist discovered four famous equations summarizing man's knowledge of electricity and magnetism. From them he obtained this equation along with another predicting the possibility of radio waves. We owe to Maxwell all our TV and radio broadcasting, our long-distance communication and radar on land, on sea, and in space. Light, X rays and other electromagnetic radiations are also governed by this fundamental equation.

## LUDWIG BOLTZMANN 1844 – 1906 $\qquad S = k \log W$

Boltzmann's equation revealed how the behavior of gases depends on the constant motion of atoms and molecules. Its great importance rests in applications in which gases play an important part: in all machines driven by steam or internal combustion; in the countless reactions among gases used by chemists to make modern drugs, plastics or other substances; in understanding weather; and even in explaining the violent processes of the sun, stars, and distant galaxies.

Plate 4

## KONSTANTIN TSIOLKOVSKY 1857–1935 $\hspace{2cm} V = V_e \ln(m_0/m_1)$

A basic part of space technology, this equation gives the changing speed of a ship as it burns the fuel that it is carrying. This equation is derived directly from one of the three great laws of motion of Sir Isaac Newton. Without it, launching space ships to the moon and the planets or orbiting the earth would be impossible; but it has also made practicable waging war with guided rockets.

## ALBERT EINSTEIN 1875–1955 $\hspace{4cm} E = mc^2$

This equation is the basis of our nuclear age. It says, simply, that a small quantity of matter can be converted into a large quantity of energy. We see this nuclear energy released in a spectacular and violent form in atomic and hydrogen bombs. But man has also controlled 'nuclear fission' in nuclear reactors that supply heat and generate electricity for our homes and factories.

## LOUIS de BROGLIE 1892–1977 $\hspace{4cm} \lambda = h/mv$

Light, a form of energy, can behave both as bullet-like particles and as continuous waves. De Broglie discovered the converse: the elementary particles of which matter is composed also have wave-like properties. His equation has had an important influence on physics, leading to modern optics, to electronic components — transistors, for example — with applications in radio, TV, computers, space ships, and military weapons. It also provides scientists with the powerful electron microscope.

forget to add: 'But, of course, for all the many mistakes of judgment in this article I alone am responsible'. Heaven help you if a reviewer calls you 'Bluffenwoofer's student' in an unfavorable review. But never thank your wife or husband who survived three years of your moodiness in a basement apartment in New Haven, living on spaghetti and climbing over stacks of spigots. Spouses should be mentioned only in a foreign language in the dedication of a book and never by name: 'À ma femme très sympathetique'.

The first footnote should include a summary of all prior work on spigots. It is important to destroy the validity of other scholarly work in order to validate your own, and to represent yourself as having read every word ever published on the subject (when, of course, you have not).

Subsequent footnotes are also the place for personal attacks on other scholars, because you need to back up your point of view in the text with facts and an argument, but not in the footnotes. There, you can say with impunity: 'I pass over the absurd theorizing of Klinkerhelften', or 'No serious scholar accepts the rather peculiar ideas of delle Aquario that spigots were smeared with rasberry juice and buried on the night of the full moon'.

Don't allow your reader to follow the footnotes easily. Constantly crisscross references: 'See footnote 19 above and also footnote 32 below'. Refer to a very important book or article in the footnotes by the abbreviation *op. cit. supra*. But cannily omit the first reference.

A certain modesty is expected of young scholars, but use your apology for not having consulted esoteric material to make yourself look even more scholarly: 'I regret my inability to consult ''Early Tibetan Faucet Spigots'' in *Folk Chronicles of Early Tibet* (1743) in the Lo-Herpa monastery because of the death of my Sherpa guide in an avalanche'.

Finally, justify your ignorance of subjects and fields other than your own by condemning all other scholarship as 'trendy', or, if it is cross-disciplinary, 'superficial': 'Dunhoffer's Pulitzer Prize-winning book, *Faucets and*

*Folklore*, is a slick and trendy overview. The pictures, however, are of some interest.'

- *The article*. If the title and the footnotes have the 'right look', almost any text will do. Probably no one will read it, but here are some general hints just in case. It is central to success to make the article as difficult to read as possible as soon as possible. One tactic is to begin with a 30- to 40-line quotation in a foreign language. German script is excellent for this purpose if your editor will agree. Another is to have the first pages contain a dozen statistical tables. Yet another is to start the discussion with a subject totally unrelated to the apparent topic suggested by your title (insofar as the reader has been able to decipher it).

A study of the chronology of Nilotic spigots could begin, 'Many Renaissance poets found in the starting and stopping of water a symbolic mode for the beginning and end of life'. It takes some pages to get back to faucets from here, and the advantage is that your reader is not only puzzled but also dimly believes that your study has some broader significance. Most readers think articles are good in proportion to their inability to understand them.

The first paragraph should also make a grandiose claim, if possible: 'A new look at Egyptian chronology and history has been long needed'. It must be written in the impersonal passive voice, however, so that you cannot be held culpable for not doing what you propose.

Style is important. Remember that the real purpose of the style you choose is to help you protect yourself. Write everything you can in the passive voice: 'When 15 Bronze Age spigots were excavated and subjected to chemical analysis, they unfortunately dissolved'. Obviously, some fool in the laboratory mixed the chemicals incorrectly, but no one can blame you. Use style to elevate the mundane to the arcane. Many words ending in 'tion' help here: 'The centralization and quantification of spigot production and distribution will be examined'.

One of the most successful ways of intimidating your reader is to invent new terminology. Such invention implies that no word currently exists to describe adequately your ideas on the subject, and may change you from a junior untenured professor to a 'seminal thinker'. So, in your discussion of spigot shapes, describe the two camps of opinion as the 'comparatists' and the 'ideospigoters'. *Never* define your terms. If possible, translate them into German.

Keep before you at all times a mental picture of the board of trustees counting papers. And never forget two important rules about scholarship:

- Its real purpose is to produce as much as possible on as little as possible.
- It is not so important for you to be right as it is for others to be wrong.

---

## The itinerant naturalist

RBP from Joel W Hedgpeth 1953 *Systematic Zoology* **2** 75.

Some journals have been devoted entirely to humor in science: *The Auklet, Brighter Biochemistry, Dopeia, Eureka, Mad Engineer, The Malpighii, The Tea Phytologist, The Journal of Irreproducible Results, The Journal of Insignificant Research,* and the *Worm Runner's Digest.* Unfortunately some have had short lives. Here is an announcement of one which was never even born.

ANNOUNCING—a NEW journal:

### THE ITINERANT NATURALIST
A journal devoted to obscure animals, obscurely published.

To be published every three months by hand on a Washington press, in 18 pt. Caslon Oldstyle; page size, 15 × 24 inches. Illustrations will consist of woodcuts, to be submitted in a finished condition by the authors.

Papers should be submitted in legible Spencerian hand (using Higgins Eternal Ink)† on ruled foolscap, not less than 50

†Genuine cuttlefish sepia also acceptable.

words and not more than 2,000 words in length, on the systematics, bionomics and unseasonal occurrence of organisms found under stones, in moist logs, moss beds, or similar biotopes, terrestrial, fresh water or marine. Descriptions of new species and notes on sexual behavior will be given priority.

Illustrations must not be larger than 2 x 3 cubits; graphs, when necessary, must not be more than one inch square.

Papers will be critically read and altered by a board of editors; proofs will not be furnished.

THE ITINERANT NATURALIST will be distributed gratis only to star route county branch libraries and remote, transmontane junior colleges.

Table of contents for Vol. I, no. 1.

Some ecological notes on a Triassic trilobite.
—*J Willoughby Fortescue*

A new genus (and new family) of stamenophilous mollusks from Equisetum.
—*Q Johannes Betulaceae*

Crepuscular periodicities in sexual vigour in an arboreal flat-worm (gen. et. sp. nov.).
—*Hieronymous Wombat*, F.L.S., F.R.S.

An unusual occurrence of Jones side-hill salmon in Deer Creek, with 15 new subspecies.
—*Anonymous*

Subscribe now, to ensure an unbroken set of this important new periodical.

Subscription rates: $15 per annum, payable in advance to Joel W Hedgpeth, Scripps Institution of Oceanography, University of California, La Jolla.

NOTE: As we do not have any italic type, italicised matter will be set in a short font of worn 6 pt. German textletter.

# A brief history of scholarly publishing

*Donald D Jackson*

From *Journal of Irreproducible Results* **16** 2, p 42 (1967). RBP of Blackwell Scientific Publications, Inc.

50000 BC  Stone Age publisher demands that all manuscripts be double-spaced, and hacked on one side of stone only.

1455  Johannes Gutenberg applies to Ford Foundation for money to buy umlauts. First subsidized publishing venture.

1483  Invention of *ibid*.

1507  First use of circumlocution.

1859  'Without whom' is used for the first time in list of acknowledgments.

1888  Martydom of Ralph Thwaites, an author who deletes 503 commas from his galleys and is stoned by a copy editor.

1897  Famous old University Press in England announces that its Urdu dictionary has been in print 400 years. Entire edition, accidentally misplaced by a shipping clerk in 1497, is found during quadricentennial inventory.

1901  First free desk copy distributed (known as Black Thursday).

1916  First successful divorce case based on failure of author to thank his wife, in the foreword of his book, for typing the manuscript.

1927  Minor official in publishing house, who suggests that his firm issue books in gay paper covers and market them through drug houses, is passed over for promotion.

1928  Early use of ambiguous rejection letter, beginning, 'While we have many good things to say about your manuscript, we feel that we are not now in position...'

1934  Bookstore sends for two copies of Gleep's *Origin of Leases* from University Press and instead receives three copies of Darwin's *Storage of Fleeces* plus half of stale peanut butter sandwich from stockroom clerk's lunch. Beginning of a famous Brentano Rebellion, resulting in temporary improvement in shipping practices.

1952    Scholarly writing begins to pay. Professor Harley Biddle's publishing contract for royalty on his book after 1,000 copies have been sold to defray printing costs. Total sales: 1,009 copies.

1961    Important case of *Dulany* vs *McDaniel*, in which judge Kelly rules to call a doctoral dissertation a nonbook is libelous *per se*.

1962    Copy editors' anthem. 'Revise or Delete' is first sung at national convention. Quarrel over hyphen in second stanza delays official acceptance.

---

# An editor writes to William Shakespeare

Henry Z Sable in *Trends in Biochemical Science* 5 XIV September 1980. RBP of Elsevier Trends Journals, Cambridge, UK.

Mr William Shakespeare
The Globe Theatre
London

Dear Mr Shakespeare,

Thank you for sending us your manuscript entitled: 'The Plays, Sonnets and Other Poems of William Shakespeare. XXXI. Macbeth.' The manuscript has been reviewed by a member of our Editorial Board, with the added advice of a referee with much experience in this area. We are happy to tell you that, in general, that manuscript is acceptable for publication, but certain changes will have to be made. In particular, the reviewers noted a tendency toward repetition, particularly in those portions in which the three witches were involved. The more pertinent and printable comments supplied by the referee and the editor are noted below.

**Act I, Scene 1.** 'Fair is foul and foul is fair.' This theme recurs in several places, and appears to be self-contradictory.

**Act I, Scene 3.** 'Thrice to thine and thrice to mine and thrice again to make up nine.' The fact that $3+3+3=9$ has been well established by many authors, and could be omitted entirely; at most it could be mentioned briefly with an appropriate citation. In the same scene each witch in turn say 'Hail'. This is one of many examples of such repetition. It should be enough to have one of them, as spokeswitch for the group, says 'Hail'. Conceivably, if union rules permit each to speak, they could do so in unison, thus avoiding any inference of seniority of one over the others. The same comments were made relative to Act IV, Scene 1, where each, in turn, says 'Show'.

**Act II, Scene 2.** At one point Macbeth states: 'Methought I heard a voice cry "sleep no more".' Clearly, there is no evidence for this statement; indeed, this whole section seems to be based upon speculation rather than fact, and should be eliminated.

**Act IV, Scene 1.** Twice in the same paragraph the three witches repeat: 'Double, double, toil and trouble, fire burn and cauldron bubble.' One of these could be eliminated with little loss of effect, and 'double, double' could, perhaps be written as '4'. On the other hand, if it is really necessary to emphasize the issue of doubling, this clearly seems to be a power series and should be expressed as '$2^n$ ($n = 1,2,3,4...$).'

In addition to these specific comments, there is some question in my mind about certain other matters. The apparent madness of Mrs Macbeth, and the repeated and lengthy conversations among Malcolm, Macduff, Ross, Siward and others in the native dialect are undoubtedly important to specialists in the field, but make the manuscript unnecessarily long. If you believe that it is essential to retain these parts of the work, the Editors request that you remove them from the body of the text and prepare them in a form suitable for direct reproduction as a miniprint supplement. We enclose a list of references to publications in which miniprint has been used, as a guide for your rewriting those portions of your play. Please note that the numbering of figures, tables, acts and scenes in the miniprint is independent of the numbering of those items in the main work.

Apart from the text, there are certain other aspects of the manuscript that need attention. The proposed title is of a form that is no longer acceptable. We have agreed to follow the standards set by the NAS-NRC Committee on Nomenclature. Consequently, we suggest that the play be entitled simply 'Macbeth'. If you wish to indicate that this is one of a series of publications on closely related topics, you should use a footnote of the form: '†This is play No. 31 in the author's series entitled "Plays".' We feel that reference to the Sonnets and Other Poems is inappropriate, since there is no question that each of those represents another, unrelated series.

Finally, the acknowledgment will have to be rewritten in more orthodox form. You have written: 'The author begs to express his most humble gratitude for the patronage of her Virginal and Most Gracious Majesty, Elizabeth, Jewel of England's Crown.' The redactory office will, undoubtedly, want to change this into the more usual form: 'Supported in part by a grant from the Elizabeth Tudor Endowment Fund.' If you would be good enough to make this change when you return the corrected manuscript, you will save a great deal of time and correspondence for everyone concerned.

Sincerely yours,
HENRY Z SABLE
p.p. Francis Bacon, for
The Editorial Board

## Authorship of scientific papers

RBP from Carl J Sindermann 1982 *Winning at the Games Scientists Play* (New York: Plenum).

The debate about correct determination of the sequence of authorship has dragged on for decades. Solutions which have been proposed include that the person responsible for the original concept and initial planning should emerge as the first author, regardless of the subsequent contributions of others to the work on which the paper is based. At a recent late-night small-group conference in a New England hotel room, the following somewhat fuzzy alternative proposals were suggested.

- If the concept and planning of the paper are developed in group discussion, first authorship can be decided by a toss of coins, with all the losers being listed alphabetically, or listed on the basis of subsequent coin tosses.
- If the research is supported by grant or contract funds, then the designated principal investigator should be the first author. If it is supported by two more or less equal funding sources assigned to different investigators, then we're back to the coin toss.
- The principles of chivalry might be involved to decide first authorship. Female scientists would be easy winners here, as would scientists of advanced age or physical frailty. (This proposal was received with a noticeable lack of enthusiasm by the group.)
- The approach of choice—elucidated just before the group was dispersed by the hotel management— turned out to be that if the research supporting a proposed paper has been started or completed without a decision about first authorship, then the sequence of authors should be determined by a complex formula, whose major elements are:

Conceptual input (C)          Hours of time invested (T)
Planning input (P)            Preparation of first draft (Pd)
Data acquisition (Dac)        Final editing (Ed).
Data analysis (Dan)

Each element is weighted, and the formula becomes

$$\frac{4C + 2P + 2Dac + 2Dan + Pd + Ed}{T^1}.$$

Assignment of values for each element is done by a select committee of peers. Differences among total scores must be tested for statistical significance, and where significance cannot be demonstrated, additional factors should be included in the formula. These are:

Years since receipt of Ph.D. (Sr)
Comparative sizes of relevant research grants (SS)
Academic rank (if university) or GS rating (if government) (R)

The formula for determination of first author provides some protection for the concept originator, but also gives strong contributions to other elements involved in the research a fighting chance at the prize. Results of initial tests of the formula with actual manuscripts can best be described as 'inconclusive.'

# Triassic dinosaur

Contributed by
David Broome.

# Delphinus Loquens?

F G Wood
*American Scientist*
**67**: 652 (1979).
RBP from
*American Scientist,*
journal of Sigma
Xi, The Scientific
Research Society.

*Behold the porpoise, with its brain*
*Large, convoluted—surely*
*Porpoises are sane.*
*And seemingly loquacious, too,*
*With raspings, barks, and whistles,*
*And Bronx-like cheers and buzzing sounds—*
*All manner of emissals.*
*The question whether they possess*
*A language such as we do*
*Is readily affirmed by some,*
*By others,* sous-entendu.
*And yet we must consider that*
*Their way of life is peaceful.*
*No internecine struggles mark*
*Their tranquil days unceaseful.*
*Their male—female relationships*
*Are always free from strife;*
*A porpoise husband never strikes*
*His girlfriend or his wife.*
*It's clear that those who think they talk*
*Are very far from right.*
*We only have to ask ourselves:*
*If they can talk why don't they fight?*

---

# Worm-breeding with tongue in cheek

## or the confessions of a scientist hoist by his own petard

James V McCon-
nell *UNESCO
Courier* **29** April
1976 p 12–15,
32.

For sixteen years now, I've published a somewhat humorous semi-scientific journal called *Worm Runner's Digest*. Herewith begins my confession. For the *Digest* started as my own personal little joke on the Scientific Establishment but has turned out to be more of a joke on me.

I've lost grants because of the *Digest*, had my laboratory experiments questioned not because of their content but because of the *Digest*, had articles I submitted to other journals turned down because I dared to cite studies published in the *Digest*. It would seem that a little humour goes a very long way—towards excommunication!

This scurrilous journal will seem humorous to you only if you happen to know a great deal about flatworms. On the off chance that some of the finer planarian points are lacking from your store-house of knowledge, I will duly explain a bit about the psychology of worms so that you will gain some insight into the psychology of worm runners. It's a strange tail, I assure you.

The planarian, or common flatworm, is a small aquatic animal that seldom grows to more than three centimetres in length and is found in ponds, streams and rivers throughout the world. I got interested in the beast because it's the simplest animal on the phylogenetic tree that has a true brain and a human-type nervous system.

But the planarian is famous for many reasons beyond its brain. For instance, it is the simplest form of life to have true bilateral symmetry—which means that you can cut it in half from head to tail and the left half will be a mirror image of the right. And it has about the most mixed-up sex life of any animal going.

Let's face it: the planarian is that psychological anomaly, an anti-Freudian animal. To begin with, it is an hermaphrodite, having a complete set of both male and female sex organs.

The flatworm lacks a mouth; instead, it has a pharynx in the middle of its body that it extrudes when it comes in contact with food. The pharynx latches on to whatever is to be the meal and the worm sucks the juices up through it as through a drinking straw.

As a graduate student at the University of Texas, I had undertaken a project with another student, Robert Thompson, to see if the planarian could be trained. Presumably, since it is the simplest animal to possess a true brain, it should be the simplest animal capable of showing true learning (or so the psychological theories of the day insisted).

So Thompson and I set up an experiment in which we

140

demonstrated—at least to our own satisfaction—that the flatworm could be taught the type of lesson that Professor Pavlov called the conditioned response. Later, when I went to the University of Michigan at Ann Arbor as a struggling young instructor, the head of the Psychology Department called me into his office for a friendly little chat.

'Jim', he told me, 'you may have heard a nasty rumour that to survive in the academic world, you must "publish or perish". I just want you to know that the rumour is true. I'm sure you know what is expected of you, but I have a favour to ask. If at all possible, will you please try to do *good* research. But if you can't, for God's sake, publish a *lot* of bad research, for the Dean won't know the difference anyhow.'

I got the message and right away set up the first 'worm lab.' at the University of Michigan.

I was given a tiny little basement room and enough funds to purchase a very modest amount of equipment and a few worms. Like all eager young instructors, I was wise enough to talk two very bright young students (Daniel Kimble and Allan Jacobson) into doing all the actual work for me.

But I had a problem; we had demonstrated that worms could learn, so what were we to do next? For a long time I puzzled over this problem, then recalled that one day when Thompson and I were working at the University of Texas, we had had a wild idea.

Planarians not only reproduce sexually, but asexually as well (one might say they have by far the best of both worlds). When a worm is first hatched from its egg, it is fully equipped to do everything but reproduce. After a few months of fattening up, it reaches puberty and begins mating. Sexual activity continues for three to four years, after which the animal seems to go into a senile decline, becomes all lumpy and misshapen—and then a miracle often occurs.

One day as the animal is crawling along the bottom of some pond, the tail develops a will of its own and grabs

141

hold of a rock and refuses to be budged. The head struggles to get things going again, but no matter how hard the head pulls, the tail remains obstinately clinging to the rock.

Not able to convince the tail to get on with it, the head does the next best thing: it pulls so hard that the whole animal comes apart at the middle. The head then wanders off, leaving the tail to manage as best it can.

Now, if you cut a human being in half across the waist, he has a tendency not to survive the operation. But if you do this to a flatworm, you merely trigger off asexual reproduction in the same way that the animal occasionally does itself. For the head will grow a new tail in a matter of five or six days, and the tail, clinging gallantly to its rock, will regenerate an entire new head (complete with brain, eyes and full sensory apparatus) in a matter of a week or two.

Furthermore, each of the regenerated portions of the beast will soon grow up to the same size as the original animal and, being rejuvenated as well as regenerated, will begin sexual mating again. (Now there's a topic that needs further research if ever I saw one!)

Thompson and I, knowing this odd habit of the flatworm, thought it might be clever if we trained a worm, then cut it in half, let the head grow a new tail and the tail grow a new head, and then tested both halves to see which half remembered the original training. Thompson and I never had the time to do that experiment at the University of Texas, but at Michigan I had students, worms and apparatus, so we set out to see what would happen.

To our great surprise, we found that the heads remembered (a month after original training) just as much as did worms that had been trained but not cut in half at all. Apparently, if you are a worm, losing your tail does not affect your memory. To our greater surprise, we found that the tails remembered even better than did the heads. Obviously, for worms, losing your head actually improves your memory!

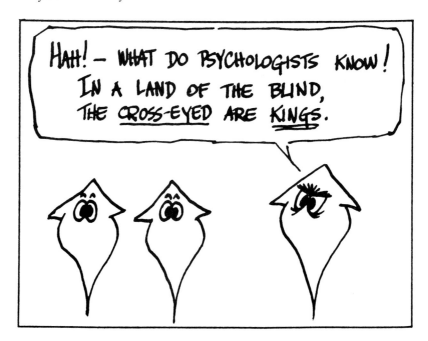

These odd results suggested to us that, in the planarian at least, memories might not be stored just in the head section. Our next experiment consisted of chopping a trained animal in several pieces and letting all of them regenerate. As we half expected, each regenerate showed memory of what the original animal had been taught.

Slowly it began to dawn on us that the usual theories of memory storage just didn't hold, for these all insisted that memories were stored neurophysiologically in the brain. Since our regenerated worms had to re-grow an entire new brain, it seemed to us that they must be storing their lessons chemically.

In other words, whenever the worm learned something, there had to be some corresponding change in the molecules in their bodies (just as there must be an electrical or mechanical change in a computer each time it stores a bit of data). Our chemical theory of memory was interesting, but how to go about proving it?

People have personalities and, after you've studied them a while, it becomes apparent that planarians do, too. That is, each organism reacts slightly differently from its cousins and brothers.

But chemical molecules are all supposed to be the same. So, when one worm learned his lesson in our training apparatus, we assumed that the chemical changes inside the body were more or less the same as those that would take place in any other worm's body when it learned the same lesson. Now, that's a perfectly tenable hypothesis if you don't happen to know much about zoology or biochemistry so, blessed with a most enthusiastic ignorance of such arcane topics, we ploughed ahead.

Here was our reasoning. Worms are rather special. Not only can you cut them in half, and each piece will regenerate into an intact organism, but you can also play all sorts of sadistic games with them.

If you slice the head in half, from the tip of the snout down to where the worm's Adam's apple would be (if it had one, which it doesn't), and then you keep the two sections of the head separated for 24 hours, each section will regenerate separately. You'll end up with a two-headed worm. Interestingly enough, a chap at Washington University in St. Louis worked with two-headed planarians later on and found, to our delight, that these animals learn significantly faster than do normal beasts. So, as far as the worm is concerned, two heads are indeed better than one!

And if two heads aren't enough for you, split each of them again, and you'll have four heads on the same body. You can get up to twelve heads at once, if you and the worm are interested in such things. More than that, you can take the head from one animal and graft it on to another—planarians don't reject tissue grafts the way that most higher organisms do.

Well, if the memory molecules were the same from one worm to another, why couldn't we train one worm, extract the chemicals from it, inject them somehow into another, and thus transfer the 'memory' from one beast to another?

For several months we tried to do just that, but we failed, simply because we were rather stupid about it all. Our hypodermic needles were far too large and we tried to inject far too much material. The poor little worms swelled up like balloons; a few popped. Eventually, though, a brainstorm hit us. Hungry planarians are cannibalistic. If we couldn't make the 'transfer' using our crude injection techniques, perhaps we could induce the worms to do the work for us.

So, in our next experiment, we trained a group of 'victim' worms and then chopped them in pieces and fed them to an unsuspecting group of hungry cannibals. After the cannibals had had a chance to digest their meal, we promptly gave them the same sort of training we had given the victims.

To our delight, the cannibals that had eaten educated victims did significantly better (right from the very first trial) than did cannibals that had eaten untrained victims. We had achieved the first inter-animal transfer of information.

After we had repeated this experiment successfully several times, we went on to show that the chemical involved in the transfer was RNA (ribonucleic acid), a giant molecule found in almost all living cells. For we showed that we could achieve this type of 'memory transfer' using a crude extract of RNA taken from the bodies of trained planarians and injected into untrained worms (using, I may say, a very, very small needle).

In recent years, a considerable controversy has cropped up concerning a whole series of similar experiments using rats and mice as subjects rather than worms. And, despite the outcries of the orthodox, it does now seem as if chemicals extracted from the brains of trained rats and injected into their untrained brethren do cause much the same sort of 'memory transfer' as we had originally discovered in flatworms.

But I stray from the point. We published our original regeneration results in 1959 and at once found ourselves mentioned in several national publications. Of course, none of the journalists took our work at all seriously but, unfortunately for us, there were hundreds of high-school students around the country who did.

Many of the brighter students in the biological sciences saw immediately that the worm could make an intriguing and most inexpensive substitute for the rat. So in 1959 we were inundated with letters from these bright youngsters asking us to tell them all about the care and training of worms. (A few more aggressive souls wrote us demanding that we send them a few hundred trained animals 'at once', because they needed them instantly and didn't have time to mess around doing the work themselves.)

I answered the first few letters personally at great length, but when several hundred arrived, it became clear that some more efficient means of communication would have to be arrived at. So my students and I sat down and wrote what was really a manual describing how to repeat the sorts of experiments we had been working on.

It took us all of fourteen pages to pour out our complete knowledge of planarianology. We typed the material up

146

and reproduced it on 'ditto' paper (using purple ink guaranteed to fade rapidly so that years later we wouldn't be embarrassed by residual displays of our youthful ignorance).

Now, I had always been noted for the oddness of my sense of humour, and the planarian research greatly enhanced this reputation. Thus none of my students considered it strange that we should try to make a joke out of this little manual, so joke it became.

First of all, it had to have a name. In psychological jargon, a person who trains rats is called a 'rat runner', because, presumably, his task is to get the rats to run through a maze or some other piece of apparatus. A man who trains insects is a 'bug runner', and someone who works with humans is, quite seriously, called a 'people runner'. Obviously we were 'worm runners', and so the title of our manual simply had to be *Worm Runner's Digest*.

One of the girls designed a crest that appeared on the cover, with a rampant two-headed worm, a coronet of connected nerve cells at the top, a Latin motto (which Arthur Koestler translates as 'When I get through explaining this to you, you will know even less than before I started'), an S and R for 'stimulus-response', a $\psi$ for psychology, and a pair of diagonal stripes painted the maize-and-blue colours of the University of Michigan. (It wasn't until years later that we learned that in the language of heraldry, diagonal stripes across your escutcheon mean that you're descended from a bastard; as I like to say, there's been a good deal of serendipity in our research.)

To top the manual off, we called it 'Volume I, No. 1', the joke being that we had no intention of continuing its publication. Little did we appreciate the strength of the publish-or-perish syndrome. Academic scientists are so desperate that they will publish *anywhere* (for the Dean really doesn't know the difference), so to our utter amazement, we began getting contributions for the next issue. Hoist with our own petard, there was nothing we could do but put out a next issue, and a next, and a next...

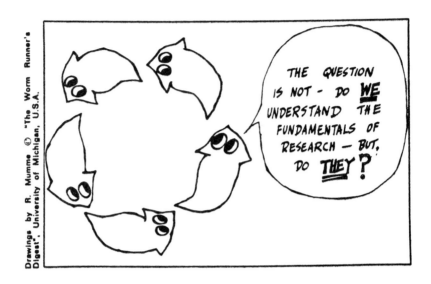

And now, here we are, a journal with 16 years behind us, an international circulation that today is numbered in the thousands. One of our crowning achievements, incidentally, was the receipt of a letter from the Library of the Academy of Sciences of the U.S.S.R. offering an official exchange of journals. We wonder still if they quite knew what they were getting.

Of course, even as our circulation increased, we remained unique. We decided that most scientific journals are deadly dull, and ours would be different. To pep things up a bit, we included poems, jokes, satires, cartoons, spoofs and short stories scattered more or less randomly among the more serious articles.

People seemed to like this melange; or at least, some of them did. A few people complained that they didn't have time to waste on the (admittedly) sophomoric humour—they wanted the 'truth' and nothing else. Their trouble was that they often found themselves getting halfway through a satire before they realized (dimly) that their leg was being pulled.

We would have ignored such complaints had they not come from some of the most famous and influential members of the scientific community. (Any conclusions you wish to draw about the qualities necessary to gather fame in the scientific community are made on your own time.)

To help these poor souls out, we resorted to a propagation device much like the worm's—namely, we split in two. We gathered all of the so-called funny stuff and banished it to the back of the journal, printing it upside down to make sure that no one would confuse the fact with the fancy.

The *Digest* inched along this way for several years, until we faced another crisis. The authors of our serious articles complained that they weren't getting adequate coverage. When an article is published in most scientific journals, it is picked up by one of the abstracting services for dissemination in abstract form.

Despite the fact that the serious side of the *Digest* contained some pretty meaty stuff, none of the abstracting services would touch anything that came from a journal with such an odd name as ours. Eventually, as a kind of last-ditch compromise, we changed the name of the front half of the *Digest*, calling it *The Journal of Biological Psychology*.

Nothing else was changed but the name, but what a difference it made! Within two months we received letters from *Psychological Abstracts, Biological Abstracts* and *Chemical Abstracts* asking that we send them this 'new' journal for abstracting. Naturally, we obliged.

As I look back at the past 16 years or so, it becomes apparent to me that life would have been a lot easier had the *Digest* suffered a stillbirth. Much of the controversy surrounding the work on 'memory transfer' stems in no small part from the fact that it received its first publication in what some of my colleagues still refer to as 'the *Playboy* of the scientific world'.

I can recall attending a meeting at Cambridge (the one in England) in 1964 at which I presented what seemed to me to be rather conclusive evidence that memories could be transferred chemically from one planarian to another. Afterwards, over the inevitable soggy cookies and warm, flavoured water (the British national drink just isn't my cup of tea), I was taken to task by a noted Scientist who informed me flatly that he would refuse to consider seriously anything published in a 'scientific cartoonbook'. When I asked him which of the British journals he had reference to, he almost dropped his cookie.

I can also remember when a good friend of mine took me aside one day to tell me how much damage I was doing to my reputation by printing the *Digest*. He was really quite worried about the matter. 'My God', he said, 'if you keep publishing articles in that thing, people may actually want to *cite* them some time, and then where will you be? You ought to change the name, throw out all that so-called funny stuff, and make it a respectable journal.'

I also treasure a letter I received from a world-famous zoologist who demanded that we remove her name from our subscription list because we were 'misleading students' into thinking that Science could be fun!

Now, in all these cases, the person doing the criticizing was a *bona fide*, expert scientist as well as Scientist. They were quite sincere in their comments, offered them up for my own betterment. I respect their scientific work, but I do feel rather sorry that so much of what is great and glorious and meaningful about science seems to have slipped through their fingers.

The kind of intropunitive wit that is the hallmark of the *Digest* can thrive only when its author is fairly secure emotionally and intellectually. People who neither understand nor appreciate humour are probably threatened by those of us who do. We speak a language they don't understand, we react to the world around us in ways that are foreign and disturbing to them. Most of them have based their entire approach to life on the premise that seriousness is next to godliness.

Those of us who see the occasional folly and ignorance of most of our (and their) behaviour often react by cracking a joke. Humour, particularly that directed against ourselves, keeps us humble in the face of our own too-well-perceived incompetence. The totally serious person fears the kind of insightful perception into his own behaviour patterns and fears humour because he cannot afford to be humble.

Now perhaps you see the *Digest* for what it really is: the house organ of an anti-Scientific movement. It is my firm conviction that most of what is wrong with Science these days can be traced to the fact that Scientists are willing to make objective and dispassionate studies of any natural phenomena at all—except their own scientific behaviour. We know considerably more about flatworms than we do about people who study flatworms. The Establishment never questions its own motives; the true humorist always does.

It is my strong hope that if we can get the younger generation to the point of being able to laugh at itself, then and only then can we hope to turn Science back into science.

**James V McConnell**

---

## 'Ginus' gives advice
**Know how to be a speaker!**

['Ginus' is the humorous newspaper of the Geological Institute of the USSR Academy of Sciences.]

Discussion of a report given by you can bring the unexpected. One can never guess beforehand what will cause an argument to leap into flames since rarely are the basic premises of the report the topic of discussion in debates. Most frequently passions boil around what hue the alluvium is colored with. Do not be confused if in discussions you should hear two contradictory opinions on one

151

question. One orator will reproach you for insufficient attention to the question of methodology; another will note that you are allotting too much space to these questions. But don't mention this contradiction: both will jump on you!

Finally do not be appalled if from a distant corner a gloomy listener arises and, having turned purple and having beaten the air with his 'horns', says that he neither understood the initial data nor the conclusions of the report. Therefore, he considers that both the report and the work of the speaker are absolutely inadequate. We advise that during the discussion you preserve a calm, business-like expression. Don't register perplexity, surprise, or perturbation. Otherwise, 'Aha, he's been caught!' the spectator will think. Very soon he will forget what is being discussed but the remembrance of your discomposure he will retain for life.

Courtesy requires that before the speaker lies a sheet of paper upon which he makes notes. It is suggested that as the final word is being prepared, usually the speaker is absent-mindedly drawing doodles. In the final words, the experienced speaker usually thanks all: both those who have spoken and those who have listened and he states that the problem has become far more understandable. After this, without any changes, he repeats the theses of the report.

---

# Four mathematical clerihews

RBP from Steven Cushing 1988 *Mathematics Magazine* **61** 23.

1. Pythagoras
   Did stagger us
   And our reason encumber
   With irrational number.

2. Kurt Gödel
   Created a hurdle
   For the truths of a system:
   You just can't list'em!

3. Wily Fermat propounded,
   'Many will be confounded
   At the thought that my theorem
   Is really quite near'em.'

4. Said Alfred Tarski:
   'Talk of "truth" is a farce. Key
   To getting it right
   Is to know "Snow is white".'

## Toward a Foreword

Vinki, Aggie, Linde Silvey and Barnabus Hughes (ed) *Mathematics and Humor*. RBP of Mathematics and Humor 1978 by The National Council of Teachers of Mathematics (excerpts).

Only people laugh. Cows may be contented, Cheshire cats may grin, and dogs and horses may play. But only men and women, girls and boys laugh. The human mind alone can see the incongruous amid the congruities of life. Unfortunately, mathematics has been lumped with war, famine, and death as something *not* to laugh about. This booklet shows that math is funny and that mathematicians laugh (even at themselves!). The Q.E.F.† is left up to you, the reader.

A few suggestions, however, may be appreciated. Since you've probably already sampled the offerings and looked at the Table of Contents, remember that the proper use of good humor demands planning. Select the items carefully; the Index at the back will be helpful. Then, there is the matter of timing—should the material be used at the beginning, middle, or end of a day's lesson? Do you want

†Here, Q.E.F. means *quod est faciendum*—'what is to be done'. In Euclidean geometry, Q.E.F. (*quod erat faciendum*) follows a construction problem and means 'which was to be done'.

the item to introduce, to lighten, or to clinch the point you are making? Perhaps you wish to offer your students something with which they can identify, such as —

A mathematician confided
That a Möbius band is one-sided
  And you'll get quite a laugh
  If you cut one in half.
For it stays in one piece when divided.

Some items may be useful on the bulletin board; others may prompt your students to create their own 'mathe*mad*ical' humor; still others may spark up a test.

At any rate, please keep in mind that if students cannot laugh while they learn, it's not *their* fault!

The MatheMADicians

\*\*\*\*\*\*

The story of Noah, his ark, and the flood is well known. What is little known is the significant role played by mathematics in the saga. As the animals disembarked,

Noah said, 'Go forth and multiply!' He found that the adders were not increasing in number. When he repeated, 'Go forth and multiply,' they replied in unison, 'We can't—we're adders'. Undaunted, Noah set his sons to work building a large table of hewn logs. This satisfied the adders, for everyone knows that adders need a log table to multiply.

* * * * * *

**Once Upon A Time** ... in an Indian village, each of three squaws prepared for the birth of her first child. The first squaw spread a bear hide near a pine grove; the second squaw carefully laid a moose hide in the shade of a large oak tree; the third squaw spread a hippopotamus hide beside a rippling brook.

It happens that the three women gave birth on the same day. The squaw on the bear hide had a son, as did the squaw on the moose hide, but the third gave birth to twin sons.

To this day, mathematicians give credit to these women for the first known proof of the Pythagorean theorem: The sons of the squaw of the hippopotamus is [*sic*] equal to the sons of the squaws of the two adjacent hides.

## Quickies

*Teacher:* Seven is an odd number. How can it be made even?
*Student:* Take away the S.

\*\*\*\*\*\*

*Archi:* Why is simplifying a fraction like powdering your nose?
*Medes:* It improves the appearance without changing the value.

\*\*\*\*\*\*

Did you hear about the probability teacher who died and went to Pair-a-Dice?
He was so *mean* that someone gave him Pois(s)on.
To make matters even worse, they called in a median to get a standard divination.

\*\*\*\*\*\*

*Pap:* If a man smashed a clock, could he be accused of killing time?
*Pus:* Not if he could prove that the clock struck first.

\*\*\*\*\*\*

*He:* Why is an hour-glass made small in the middle?
*Ro:* I don't know, unless it's to show the waist of time.

\*\*\*\*\*\*

*Tom:* Wasn't Isosceles the Greek philosopher who died of hemlock poisoning?

\*\*\*\*\*\*

The polite mathematician says, 'You go to infinity—and don't hurry back!'

\*\*\*\*\*\*

157

The mantissa is the dismal part of a logarithm. (Found on a trigonometry examination paper.)

\*\*\*\*\*\*

Show me two cars enmeshed after an auto accident and I'll show you a rectangle (wrecked tangle). (*Fred Pence*)

\*\*\*\*\*\*

Show me a barbershop quartet convention and I'll show you a harmonic function. (*Inspired by Charles W. Trigg*)

## Miniprofiles

*Archimedes, stick in hand,*
*Traced his tombstone in the sand.*
*Khayyam laid cubics on the line—*
*But better known for a jug of wine.*

*Fibonacci couldn't sleep—*
*Counted rabbits instead of sheep.*
*Fermat found margins a handy place,*
*But all too soon ran out of space.*
*Evariste Galois fought a duel—*
*Fate was ruthless, fate was cruel.*

*Hamilton crossed a Dublin bridge—*
*Carved graffiti on its ridge.*
*Emmy Noether—Adam's rib—*
*Antedating women's lib.*
*Gödel—giant-stride agility—*
*Decided undecidability.*
*Bourbaki keeping fit and nifty—*
*Component parts retire at fifty.*

*Katharine O'Brien*

---

Contributed by
Frank Vastola.

REVIEW

This paper should be greatly reduced or completely oxidized.

Attributed to
Deming by B R
Bertramson.

REFLECTION

It is disconcerting to reflect on the number of students we have flunked in chemistry for not knowing what we later found to be untrue.

A R Amundson.

RELIABILITY

There is a fair body of chemists who hold that you must not believe any data that are not confirmed by theory.

---

# The love-lorn chemist—an enigmatic epic

*B J Hazard*

*'I love you!' he cried as he filled his pipette,*
*With a love that is not merely formal.*

*I have loved from afar since the day we first met*
*With love stronger than twice N/10.*                    'decinormal'

*'The world-shaking powers of atomic U*                  'uranium'
*And the similar powers of Pu*                           'P — U'
*Have not the effect on my unstable cranium*
*Of the slightest of glances from you.*

*'In °F above Helen of Troy*                             degrees
*(Or Helen of avoirdupois?)*                             Fahrenheit
*You're the pure metal without the alloy,*              —'degrees far in
*The reagent of grade 'Analar'.*                         height'

*'Donna é mobile, but you have great P*                  'poise'
*Unmeasured by any viscometer;*
*Your feminine charms exceed Myrna Loy's,*
*So now from my thoughts I'll omit her!*

*'Though the earth below and the moon above*
*Are as different as CaCO$_3$ and cheese,*               'chalk'
*As constant as either will be my love*
*Through endless ∞'s.*                                   'infinities'

*'I will love you still when you are old—*
*And your Ag-grey hair will I kiss—*                     'silver-grey'
*With a love that's as pure as thrice-refined Au*        'gold'
*And as fierce as a Skraup synthesis.*

*'Oh, tell me you are not neutral,' he said,*            barium—'bear I
*''T would be too hard to Ba sure,*                      am'
*My heart would become quite as heavy as Pb*             'lead'
*And I'd Zn in a swoon to the floor.*                    zinc—'sink'

*'As this* 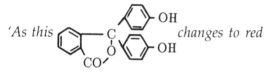 *changes to red*         'phenophthalein'

*When the neutralisation's completed,*
*The blush in your cheeks tells me that I have said*
*Enough to make you quite heated.*

160

'Oh! Lean on me Annie, oh! $do$ 
Lay your head on my chemical chest;
You're as light as a feather, as light as $H_2$
And fill me with heavenly zest.

'Come raise your hydrostatic head,
You need not fear seduction;
Bring close to mine your $[-C_6H_4-N=N-C_{10}H_5(NH_2)(SO_3H)]_2$
Lips—I'll apply some suction!

My whole visible absorption's in you,' he said,
But—though I am no prig—
In V you'll be till we're out of the R—
Beyond that would be infra—dig!

'Although I'm precipitate, I need no love filter

Or $CH_2OH$, B, or C—

I assure you, my darling, that I am no jilter—
I'm 100% affinity.

'C is my attraction,
Not W-ing or in spurts;
My rate of osculation
'S so high it almost Hz.

'Gone is all my
My circuits—and pants—are shorts;
My in this instance
Comes not in pints but $SiO_{2(hex)}$

'Come, live with me and we will laugh
At what the canteen serves.
We'll start a joint production graph
And I will draw your curves.

Oh! fly with me to much cooler climes
Where the ambient temperature's lower,
As $I_2$ does when it sublimes,
Or S beginning to flower.

aniline—'Annie lean'

'H-two'

'Congo Red'

violet—
'(in)violate'
'red'

'philtre'

'vitamin A'

'one hundred per cent'

'constant'
Weber—'waver(ing)'

Hertz—'hurts'

'resistance'

'capacity'
quartz—'quarts'

'iodine'
'sulphur'

# The love-lorn chemist—an enigmatic epic

'On my new Carnot cycle we'll ride to the ocean—
I've lots of F;                                        'free energy'
I've discovered a source of perpetual emotion
And decreased my S!                                    'entropy'


'So W about making an Ω—No, don't howl,
In a V I have current-cy A-le.
Not F we'll dyne cheek by J
And alternate with H.T., e.g.

watt—'what';
ohm—'(h)ome'
volt—'vault';
amp—'ample'
faraday—'for a
day'; joule—
'jowl'; high
t(ension)—'high
tea', for exam-
ple

'Oh! We will live a happy life
And all quarrels eliminate;
We'll eschew ev'ry kind of strife
And thoughts $HgCl_2$.                                 'corrosive
                                                        sublimate'

'If you will be my graphical x,
Then I will be your y,
Entirely co-ordinate in all best respects,
Dependently variable—I!

We'll $\int_A^B$                                        'integrate from A
The functions of our sexes                              to B'
And perhaps in the fullness of time we shall see
Some little dy/dx's                                     'dee why by dee
                                                        ex's'

'Unless you tell me you love me yourself
Tears will flow and will have to be dried.'
E'en as he said this he reached to the shelf
And brought down the $CaCl_2$.                          'calcium chloride'


'I must say,' he cried, shaking up his emulsion,
'To inspire you with love I have tried
And yet you experience only revulsion—
Do I smell like $(CH_3)_2S$?'                           'dimethyl
                                                        sulphide'

162

*He appealed and entreated with eyes all aglow,*

anisole—'Annie's sole'

*But* ⬡—OCH₃ *response was flaccid*

'H-two-O'

*So he drank deep—not H₂O*
*But conc. HCN*

'concentrated prussic acid'

*The moral of this sad history*
*Is that pleasure and work do not mix;*
*The most violent reactions in chemistry*
*Are those of the opposite sex!*

---

## The chemical engineer

*Anon*

A Chemical Engineer is a man who knows
    Enough Engineering to confuse the Chemist
    Enough Chemistry to confuse the Draftsman
    Enough Mathematics to confuse his boss
    And enough Electricity to confuse himself.

His world consists of a series of simplifying assumptions and derived basic equations.

All of which he knows are true because they can be integrated twice and sometimes thrice.

---

## Chemical structure puns

Contributed by Alexander G Briggs

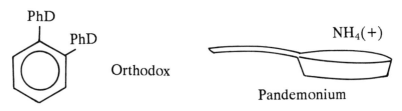

Orthodox

Pandemonium

163

# Chemical structure puns

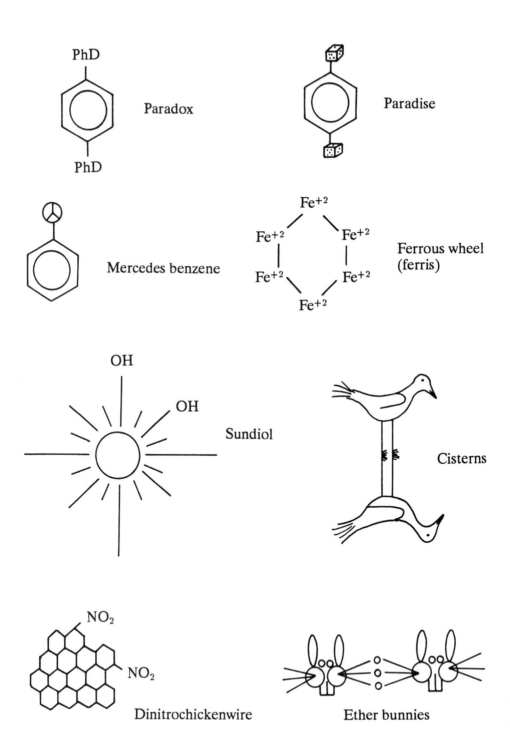

Paradox

Paradise

Mercedes benzene

Ferrous wheel
(ferris)

Sundiol

Cisterns

Dinitrochickenwire

Ether bunnies

Transparent

Cis-co-kid

## Truths, half-truths, and statistics

RBP from John Allen Paulos 1985 *I Think, Therefore I Laugh: An Alternative Approach to Philosophy* (New York: Columbia University Press) p 78–82

Benjamin Disraeli coined the phrase 'Lies, damn lies, and statistics' and the phrase as well as the sentiment has lasted, though I like 'Truths, half-truths, and statistics' better. In any case even relatively simple applications of statistics can cause problems, not to mention the horrors associated with things like the often misinterpreted SPSS computer software (Statistical Programs for the Social Sciences).

Probability and statistics, like geometry and mathematics in general, come in two flavors, pure and applied. Pure probability theory is a formal calculus whose primitive terms are uninterpreted and whose axioms are neither true nor false. These axioms originally arise from and are made meaningful by real-life interpretations of terms like 'probability', 'event', and 'random sample'. The problem with applying probability and statistics is often not in the formal mathematical manipulations themselves, but in the appropriateness of the application, the validity of the interpretation, and indeed the 'reasonableness' of the whole enterprise. This latter activity goes beyond mathematics into the sometimes murky realm of common sense and the philosophy of science (grue-bleen, ravens, etc.). Even though 1 and 1 equals 2, one glass of water and one glass of popcorn does not equal two glasses of mixture. The mathematics is fine, the application is not.

\*\*\*\*\*\*

165

*Martha:* What did you get for the density of the block, George?

*George:* Well, it weighed about 17 pounds and had a volume of about 29 cubic feet, so I guess the density is 0.58620689551 pounds per cubic foot. This calculator's really swell.

\*\*\*\*\*\*

Babe Ruth and Lou Gehrig played baseball for the New York Yankees. Suppose Ruth had a higher batting average than Gehrig for the first half of the season. Suppose further that during the second half of the season Ruth continued to hit for a higher batting average than Gehrig. Is it nevertheless possible for Gehrig's batting average for the entire season to be higher than Ruth's batting average for the entire season? The fact that I've used up a paragraph asking the question indicates that the answer is yes, but how can it be?

One way it can be is for Ruth during the first half of the season to hit for an average of 0.344 getting 55 hits in 160 times at bat while Gehrig during this same time hits for an average of 0.341 getting 82 hits for 240 times at bat. During the second half of the season Ruth's average is 0.250 since he gets 60 hits in 240 times at bat whereas Gehrig's is 0.238 as he gets 38 hits in 160 times at bat. Nevertheless for the season as a whole Gehrig's average of 0.300 is higher than Ruth's average of 0.287.

Thus even the third grade notion of an average can be misused, not to mention (as I already have) things like complicated multidimensional analyses of variance.

\*\*\*\*\*\*

If Waldo comes from country *x*, 30 percent of whose citizens have a certain characteristic, then if we know nothing else about Waldo, it seems reasonable to assume that there is a 30 percent probability that Waldo shares this characteristic. If we later discover that Waldo belongs to a certain ethnic group 80 percent of whose members in the region comprising countries $x$, $y$, and $z$ have the characteristic in question, what now are Waldo's chances of sharing this characteristic? What if we subsequently determine that Waldo belongs to a nation-$x$-wide organization only 15 percent of whose members have this characteristic? What now can we conclude, with all this information, about Waldo's chances of having the characteristic in question?

\*\*\*\*\*\*

Wittgenstein writes about a man who, not being certain of an item he reads in the newspaper, buys 100 copies of the paper to reassure himself of its truth. Given the extent to which newspeople and the media report each other's reports, checking in several different papers or periodicals is much more intelligent.

\*\*\*\*\*\*

*News release:* Abortions are becoming so popular in some countries that the waiting time to get one is lengthening rapidly. Experts predict that at this rate there will soon be a one-year wait to get an abortion.

Projection of 'trends' linearly into the future is often about as reliable as this 'news release.'

\*\*\*\*\*\*

Most automobile accidents occur close to home, so we can see that near one's home is the most dangerous place to drive.

Very few accidents occur when one is driving over 95 mph, so it's clear that driving this fast is actually quite safe.

\*\*\*\*\*\*

It is a surprising, almost counterintuitive, fact that if just 23 people are chosen at random from a telephone directory, the probability is 0.5 (chances are about 50 – 50) that at least two of them will have the same birthday. Recently someone was trying to explain this oddity on a television talk show. The incredulous host thought the man must be wrong and so asked the studio audience how many people had the same birthday as he did, say March 19. When nobody in the audience of about 150 people responded, the host felt vindicated and the guest felt embarrassed. Actually the question the host had raised was very different than the one the guest had been discussing. It turns out that a randomly selected group of 253 people is required in order for the probability to be 0.5 that at least one member of the group has any *specific* birthdate (such as March 19); a group of only 23 people is required for the probability to be 0.5 that there is *some* birthdate in common.

This is a specific instance of a very general phenomenon. Even though any particular event of a certain sort may be quite rare, that some event of that sort will occur is not rare at all. The American science writer Martin Gardner illustrates this point with the story of a spinner which is equally likely to stop at any of the 26 letters of the alphabet. If the spinner is twirled 100 times and the results recorded, the probability of any *particular* three-letter word, say 'cat', appearing is quite small whereas the probability of *some* 3 letter word appearing is very high.

Columbus discovered the New World in 1492 while his fellow Italian Enrico Fermi discovered the atomic world in 1942. John Kennedy, elected President in 1960 and assassinated in office, had a Lincoln for a secretary while Abraham Lincoln, elected President in 1860 and assassinated in office, had a Kennedy for a secretary. As Gardner has noted, the acronym formed by the planets listed in order—Mercury, Venus, Earth, Mars, Jupiter, Saturn, Uranus, Neptune, Pluto—is M V E M J **S U N** P, while that for the months is J F M A M J **J A S O N** D. In each case we have an unlikely happening whose type, while almost impossible to specify precisely, is quite likely to have some instances. The relevance of this to evolution

168

should be clear. That a particular branch should have evolved as it has is quite improbable. That some such branch should have evolved is much less so.

******

The probability of getting at least one head on two flips of a coin is 0.75. The chances of rain tomorrow are 75 percent. I think the odds of George's marrying Martha are 3 to 1. Does 'probability = 0.75' mean the same thing in each of the above cases? The preceding discussion concerning unlikely coincidences might be summarized aphoristically as, 'It's very improbable that no improbable event will occur.' Are both uses of improbable in this statement the same?

IT IS A CLASSICAL IMPOSSIBILITY TO LIVE UP TO THE EXPECTATIONS OF A POST—DIRAC WORLD!

# Mystery, Magic and Medicine

Faith and knowledge lean largely upon each other in the practice of medicine.

*Peter Mere Latham*

# The language of medicine

John H Felts in
*Verbatim* VIII no 4
Spring 1982 p 12
(abbreviated). ©
Copyright 1982
by VERBATIM,
The Language
Quarterly. Used
by permission.

The requirements for hospital accreditation in the United States are such that house officers and staff physicians must be compulsively complete in recording what happens to patients. As Dr McArdle has indicated, this is something of a chore at which most of us are less than competent. Medical college curricula are so laden with courses in the sciences, basic and behavioral, that proficiency in our language can hardly be expected. In 1898, the University of North Carolina School of Medicine asked only that matriculants pass examinations determining their proficiency in English composition, arithmetic, algebra, and Latin; today the requisite courses for entry into most American medical schools are chemistry, biology, and physics. The need for brevity, simplicity, and clarity in medical writing is still recognized and appreciated, but only the most ardent optimist has any hope for improvement in our prose anytime soon.

Most hospital record rooms have their own selection of amazing arrangements of words put together by dictator and the typing pool. Here are some, for which doctors, I fear, must assume major responsibility.

- The left leg became numb at times and she walked it off.
- The patient has chest pain if she lies on her left side for over a year.
- Father died in his 90s of female trouble in his prostate and kidneys.
- Both the patient and the nurse herself reported passing flatus.
- Skin—somewhat pale but present.
- On the second day the knee was better, and on the third day it had completely disappeared.
- The pelvic examination will be done later on the floor.
- By the time she was admitted to the hospital her rapid heart had stopped and she was feeling much better.
- If he squeezes the back of his neck for four or five years it comes and goes.

- Patient was seen in consultation by Dr—, who felt we should sit tight on the abdomen, and I agreed.
- Dr—is watching his prostate.
- Discharge status: Alive but without permission.
- Coming from Detroit, Mich., this man has no children.
- At the time of onset of pregnancy the mother was undergoing bronchoscopy.
- Healthy appearing decrepit 69-year-old white female, mentally alert but forgetful.
- When you pin him down, he has some slowing of the stream.

---

## Humor plus humility equals humaneness

*Harold Stevens*

Extracts from Humor and humility equals humaneness, *JAMA* 190.13, 28 December 1964 pp 88–91. Copyright 1964, American Medical Association. Reproduced with permission.

If you have any hope of coping with 'system', you will have to arm yourself with a well-developed sense of humor.

This indispensable quality is certainly not taught in medical school. Perhaps the faculty should have included it in the new curriculum. I must now confess that I tried, but my efforts were either too clumsy or too subtle. In a modest way I tried to develop a sense of humor by introducing you to one of the most sensitive, perceptive, and articulate students of human behavior in our time, James Thurber. He should be required reading along with many others who have been blessed with some of his combination of insight, sensitivity, compassion, and understanding.

One of the keenest depictions of human frailty is Thurber's brief but eloquent and classic story, *The Private Life of Walter Mitty*. You will recall that the timid Walter Mitty sees himself in fantasy as a great doctor who casually walks into the operating room to find the surgeon desperately struggling with insurmountable problems.

Not the least of these is a broken anesthesia machine which Mitty promptly, dexterously, and with quiet efficiency repairs from parts of his own fountain pen. But other catastrophies occur and, in a final gesture of complete demoralization, the surgeon cries out, 'My God, coreopsis has set in.' Walter Mitty quickly steps up to the table, analyses the problem, and with a cool eye and steady hand, dexterously completes the operation and saves the patient.

It was my hope, by using subliminal influence worthy of Madison Avenue, to refresh your memory of this Thurber classic. Accordingly I worked a reference to it into the final exam in your second year. Perhaps you recall this multiple-choice question:

> An 8-month-old infant suffered *Hemophilas influenzae* meningitis. Antibiotic treatment was started at once. In one week his fever subsided. But in two weeks he was irritable, vomited, and low-grade fever appeared. He became lethargic and refused his bottle. The neurological examination was otherwise negative. The spinal fluid was normal four days after the above antibiotic treatment was started, and remained so. His trouble is:
> a. subdural effusion
> b. post-meningitic behavior disorder
> c. coreopsis has set in
> d. recurrent meningitis
> e. hydrocephalus.

I did not expect any loud guffaws but I did expect at least a few smiles and some recall of the heroics of Walter Mitty. To my dismay, no one in the class recognized the phrase. No one realized that coreopsis is not a disease but a flower. What's more, seven students marked this as the correct answer. I was crushed. Later, I reported this to a fellow-admirer of Thurber, Mr Edward Newhouse, a writer for *The New Yorker* magazine. He asked me for a copy of the examination. With the examination I enclosed a cover letter expressing my dismay and disappointment not that seven students got the question wrong, but that they did not know Thurber. Mr Newhouse sent the copy of the exam to Mr James Thurber and here is Thurber's reply:

Dear Eddie:

I wish I knew Dr Stevens for he is one of my favorite people. I frowned through my laughter about those 7, not because they didn't know Thurber, but because they did not know what was the matter with the baby. Maybe it is because I am now the grandfather of a girl of 4 and a boy of 2. Incidentally, I happened to meet the late James Gerard in Nassau 10 years ago, a wonderful guy who had a couple of drinks with us each evening on the ship coming home, then invariably got up and said, coreopsis has set in, and went to his stateroom. It is my firm but sad conviction that ninety percent of Americans would figure coreopsis is a disease and I can only hope that the women's eyes do not brighten with the hope of flowers when a man says, 'I've brought you syphilis.'

# What's funny about doctors?

*Walter Blair*

From *Perspectives in Biology and Medicine,* Autumn 1977 pp 89–98. RBP of Walter Blair.

[Adapted from an address delivered to the American College of Physicians Associates, Chicago, June 1976.]

Since laughter is a subjective reaction, different groups are sure to disagree concerning what is funny about doctors. Patients, for instance, who feel that physicians have peculiar (not to say inconvenient) ideas about their life-styles, can't believe that physicians know; and physicians badmouth the notions of patients. I'll therefore avoid evaluations and merely report what humorous writers, practically none with medical training, have indicated they believed.

A striking fact is that, from ancient Grecian days to the present, most comic picturings of doctors have been quite similar.

Those wise old Greeks who instructed the world about so many matters discovered, defined, and portrayed two everlastingly comic types, and humorists identified medical men with one of them.

One type, the *eiron*, pretends to be less knowing or capable than he actually is. An immortal joke goes this way: A self-effacing, uneducated, unimpressive fellow turns up. Some smarty—or several smarties—decide that he isn't much and try to put him down. The ironic man outsmarts his detractors. Or someone who looks abysmally stupid proves that he's no such thing by coining wise and witty sayings that are quoted for centuries. Socrates, if Plato is trustworthy, was an ancient example. Cervantes immortalized the type when he created earthy Sancho Panza. (Sancho, who said, 'I'm a fool', also said, 'The proof of the pudding's in the eating', 'All is not gold that glitters', and 'Honesty's the best policy.') In our country, Franklin's Poor Richard was an early bird of this sort. (He also said, 'Honesty's the best policy', and he made other profound remarks, such as 'Let thy maid-servant be faithful, strong, and homely', and 'There are more old drunkards than old doctors.') The type has been a great favorite in the United States from Ben Franklin's day through the days of Hosea Biglow, Josh Billings, and Will Rogers, down to the era of Senator Sam Ervin, Ann Landers, and Erma Bombeck.

For some reason, humorists during several centuries didn't tend to think that doctors belonged to the self-deprecatory but canny group. Instead, they made them members of a second comic class—the opposite of the *eirons*. Greeks called their ilk *alazons*. A modern critic defined the *alazon* thus: 'A deceiving or self-deceived character, an imposer, someone who pretends or tries to be something more than he is.' Aristophanes put a specimen into *The Frogs*, and Plautus used versions in seven comedies. Claiming 'to be more than he is' in a number of ways, the bluffing braggart in several guises got laughs down through the ages.

In one popular incarnation, the type was a soldier who boasted about his bravery until he faced the test, then quickly showed he was a farcical coward. (Shakespeare's Falstaff was the most famous example.) Or the *alazon* put on a show of knowing more than he actually did know—and this was the role which comic writers constantly assigned to doctors.

Look at the type as he was pictured by a very popular

group of actors from the middle of the sixteenth century to the middle of the eighteenth century—that long era's equivalent to today's stage, movie, comic strip, and television idols combined. For actors of the *commedia dell' arte*—'the comedy of experts'—charmed not only Italian audiences at home but also British, French, Austrian, German, Polish, and even Russian audiences away from home; and their comedy strongly influenced that of Molière, Jonson, and Shakespeare.

Every one of the many companies had a specialist who throughout his career played the stereotyped role of 'the Doctor'—'practically the same character', as a scholar says, 'as had been portrayed in the preceding centuries.' That means that he was prissy, pompous, and pedantic, and that he let on that he was a jack-of-all-scholarly-trades who knew everything. He impressed less-learned people by flinging around foreign words and phrases or sesquipedalian coinages that he hoped would sound impressive. The players didn't specify what the fellow was a doctor *of*—medicine, of course, but also (if he was to be believed) of philosophy, astronomy, even (I don't know how this happened) of literature.

Thanks in part to the influence of the Italian actors and the dialogues that they concocted, the same characterization was often repeated in England. A whole book has been written about doctors in Elizabethan plays. The picture hardly ever varies, nor does the way these men talk; always they use big words recklessly to hide the fact that they don't know what they're talking about. One nondoctor in a play advised a friend: 'Beat not your brains to understand their parcel-Greek, parcel-Latin gibberish.'

On and on through the years, the libelous representations continued. Even a number of doctors who happened to be witty writers kept up the assault and battery—Rabelais in France, Tobias Smollett and Oliver Goldsmith in England, and Oliver Wendell Holmes, Sr, in the United States.

A couple of great dramatists who spent large parts of their lives kidding doctors were outstanding among the hosts of literary men in the long parade. France's greatest dramatist, Molière (1622–1673), and one of England's greatest, George Bernard Shaw (1856–1950), though they were two centuries apart in time, were alike in constantly using medics for comedy. And what they had to say got attention and had, it may be, some value.

The French playwright wasn't off the mark when he had a character say, 'Your Molière is an impudent fellow with his comedies, and I think he oughtn't put such honest men as doctors on the stage. ... If I were one of them, (and) if ever he fell ill, I'd let him die without professional help.' In six plays, Molière brought up all the old charges plus a few new ones that he'd thought up himself.

One of his characters trots out an age-old accusation: 'Most doctors have much classical learning. They speak good Latin. They can name all the diseases in Greek, define and classify them. But they know nothing about curing.' A doctor in another play flings around the customary batch of high-sounding words:

> First, to cure this obdurate plethora and this luxuriant cacochymy throughout the body, I am of the opinion that he should be liberally phlebotomised; that is to say he should be bled frequently and copiously, first, at the basilic vein, then at the cephalic vein, and ... the vein in the forehead. ... At the same time he should be purged deobstructed, and evacuated by proper, suitable purgatives, that is, by cholagogues, melangogues, et cetera, for since the source of all the evil is either a gross and feculent humor, or a black and thick vapor which obscures, infects and contaminates the animal spirits, it is proper that he should afterwards take a bath of soft, clean water with plenty of whey.

Rather newer were Molière's jokes about the way physicians in his day habitually tended to believe ancient authorities even when they had been proved wrong, and about professional etiquette. A doctor is told that a patient has died. 'Impossible!' he says. 'You must be mistaken.' 'Well,' he is told, 'he's dead and buried. I saw it myself.' 'Quite out of the question,' snaps the doctor. 'Hippocrates says that such maladies last either fourteen or twenty-one days, and it's only six days since he fell ill.'

Another doctor praises himself for following strict rules

regardless of a patient's suffering: 'I'm devilish strict about it, except among friends. When three of us were called in consultation the other day with an outside physician, I stopped everything—wouldn't let anyone say a thing until matters were conducted according to rule. People in the house did all they could—patient died bravely—but I wouldn't give way.'

Doctors and medicine were interests that absorbed George Bernard Shaw during a great many of his 94 years. Roger Boxill in 1969 published a 200-page book crammed with information about these interests and summarized: 'Nearly two dozen doctors appeared as characters in his plays and novels; some of the most important medical men of the day were among his acquaintances ... and if all his medical articles, letters and speeches were collected, they would fill a rather substantial volume.'

Boxill finds that 'the Shavian comedy of medicine concentrates on exactly the same target as does the Molièrean.' Among them are such antiques as bragging, the flinging around of long technicological Latinate terms. The claims and the pretenses were lamentably incongruous, Shaw kept pointing out, with the inability of the braggarts to effect cures.

Shaw also took humorous or satirical jabs at the late-nineteenth and early-twentieth century equivalents to flaws that Molière had joked about. Molière's physicians reasoned their way to conclusions on the basis of premises furnished by Claudius Galen and Hippocrates—scholastic theories that often led those who automatically followed their teachings to unsound conclusions. Shaw's doctors base their deductions on the theories of modern medical scientists. One doctor, Sir Ralph Bloomfield, figures that the germ theory is the key. 'Find the germ of the disease'; he says, 'prepare for it a suitable anti-toxin; inject it three times a day ... phagocytes are stimulated; they devour the disease; and the patient recovers—unless, of course, he's too far gone.' A surgeon, Cutler Walpole, has decided that all illnesses are brought about in another way, and he knows how to deal with the lot of them: 'I tell you, blood-poisoning. Ninety-five percent of the human race suffer from chronic blood-poisoning and die of it. It's as simple as A.B.C. Your nuciform sac is full of decaying matter. ... Let me cut it out for you. You'll be another man.'

Shaw's second target—his period's professional etiquette—is the professionalism that keeps physicians from disagreeing with one another. Dr Bonington knows that the cure-all that Dr Walpole champions is no such thing, but he says: 'The nuciform sac is utter nonsense; there's no such organ. It's a mere accidental kink in the membrane, occurring in perhaps two and a half percent of the population. Of course I'm glad for Walpole's sake that the operation is fashionable; for he's a dear good fellow.' Dr Ridgeon is just as sure that phagocytes haven't anything to do with the resistance to disease provided by vaccine therapy, but professionalism leads him to keep his mouth shut.

Despite the fact that Shaw had some quite loopy ideas about medical men and medicine that he worked into silly scenes, several leading physicians agreed that (to quote one of them) in his writings 'there is a great deal of truth to be found, and [it is] expressed with a great deal of gentleness towards our profession.'

During the 3¼ centuries between Molière's birth and Shaw's death, the New World produced writers who joined Old World writers in kidding overlearned doctors. If anything, our countrymen were likely to have more fun than Europeans doing this because they felt that scientists, professors, physicians, and their like affronted common men who possessed the most precious of all gifts—gumption. (Tom Paine could give the Declaration of Independence a powerful boost by writing a pamphlet called *Common Sense*, and even in the nation's early days some politicians found that they got votes when they deliberately downgraded their grammar.) Washington Irving constantly jested about 'natural philosophers' who reasoned their way to outrageous conclusions and (as he put it) 'maintained that black was white, and that, of course, there was no such color as white.' One of James

Fenimore Cooper's favorite targets was the pretentious pendant.

Herman Melville, who had a flair for humor and satire, in his demi-autobiographical novel, *White-Jacket*, about life aboard a man-of-war, wrote a memorable bit concerning a doctor. The very look of Cadwallader Cuticle, M.D., surgeon of the fleet—'withered, shrunken, one-eyed, toothless, hairless'—betokens spiritual and mental decay. He consults fellow surgeons, disregards their advice, then before an audience of students and colleagues performs what he modestly calls 'an excellent operation.'

He pours out talk in a style endlessly assigned to fictional doctors, flaunting technical terms, dropping names, citing and refuting or approving authorities. All this tortures a pain-wracked patient who of course in those times isn't anesthetized. The end of the scene:

> 'I must leave you now, gentlemen'—bowing. 'To-morrow, at ten, the limb will be upon the table, and I shall be happy to see you all upon the occasion. Who's there?' ...
> 'Please, sir', said the Steward, entering, 'the patient is dead.'
> 'The body also, gentlemen, at ten precisely', said Cuticle. ... 'I predicted the operation might prove fatal; he was much run down. Good-morning'; and Cuticle departed.

Catering to a lower-browed audience, a great many American humorists—the sort with a notorious liking for wild incongruities—helped their fellows think better of doctors by writing about some men who completely outdid them in boastful eloquence. These were the mountebanks who thrived on the farms and the frontiers. Traveling around the country, they staged medical shows, peddled worthless cures, and (if they were lucky) got away before sheriffs or lynch mobs caught up with them. Here's a fifth of the spiel of one of these hawkers in 1851 on the Alabama frontier. He's talking up 'the all-healing, never-failing, spot-removing, beauty-restoring, health-giving e-ra-sive Soap':

> Why, gentlemen, when I first became acquainted with this inextollable gift of divine Providence, I had an obstruction of the vocal organs, an impediment of speech that bid fair to destroy the hopes of the fond parents who intended me for the bar or the

pulpit. I was *tongue-tied*—but I came across the precious compound—swallowed just half an ounce, and ever since ... I have been volubly, rapidly, and successfully, ... sounding the praises of the incomparable, infallible, inimitable, ... magical, radical, tragical, erasive soap!

Robert Louis Taylor, 110 years after the soap-seller's spiel, published *Journey to Matecumbe*, a novel so clearly patterned after *Huckleberry Finn* that even some reviews noticed the resemblance. The boy narrator tells about the performance of 'Dr' Ewing T Snodgrass in a hall that he hires in Boskey Dell, a little town on the Mississippi river. He is offering the crowd that gathers his 'celebrated supportatory trusses, constructed after a design by the hundred-and-forty-year-old Choctaw Chief Wah-Wah-Too-Se':

> '*I was totally unable to walk till I was twenty-four years old!* ... I'd had a compound double hernia ... with complications of fibrosis, osteo-jaundice, and yaws. ... I was running, leaping streams ... within a week after I put these trusses on.'
>
> Dr Snodgrass then told the crowd a little about hernia, and I was glad to hear it, because I'd never exactly realized what it was, before. He explained how the muscles in the outside stomach, or peritonitis, got inflamed from eating the wrong kind of food, and caused an inner swelling not visible except under a microscope, where the legs joined onto the body part. The sockets don't heave and haul right. At the least, a person felt tired now and then, and ... at the worst, your legs fell off.
>
> You could see it was a satisfaction to the people to hear about this, because it was entirely new information to most of them, just as it was to me.

Since speeches that comic writers invented for fictitious doctors were caricatures of real doctors' utterances, the outrageous nonsense of charlatans such as the soap man and Snodgrass were caricatures of caricatures—parodies which made real doctors sound like paragons of sanity.

In the field of genteel humor, one Boston doctor who was also a writer, and a charming and witty one at that, encouraged readers to laugh *with* a medical man instead of *at* him—Oliver Wendell Homes. He did this during a large part of the nineteenth century because he was long-lived (1809 – 1894), he started publishing early, and he kept at it until his final years.

Everybody knew that jolly little Holmes had started his practice by announcing 'small fevers gratefully accepted', had made real contributions to knowledge, and for years was a beloved teacher in Harvard's medical school. In poems such as 'The Stethescope Song', he invited laughter at his Latinate technical vocabulary and his own fallibility; in his popular 'Breakfast Table' series he constantly took potshots of the old-fashioned sort at physicians; and in a series of 'medicated novels', he pictured physicians who (like Holmes) were sensible and witty. Holmes's most famous poem, 'The One-Hoss Shay', was about 'a deacon's masterpiece' built so logically that every part was equally strong and only after a century,

> *You see, of course, if you're not a dunce,*
> *How it went to pieces all at once. ...*
> *End of the wonderful one-hoss shay,*
> *Logic is logic. That's all I say.*

The doctor who wrote that wasn't at all likely to depend too thoroughly, as Shaw's physicians do, upon deductive reasoning. What's more, when *a doctor* kidded *doctors*, including himself, *he* became an *eiron*.

Skilled—though generally unfunny—writers did a great deal to create sympathetic images of medical men and thus discredit satirical representations. Bret Harte, who probably was better at popularizing stereotypes than any other of our storytellers, in 13 short stories and a novel made readers thoroughly familiar with his lovable Dr Duchesne, an intelligent, skilled, brave mining-camp practitioner who was as famous for his compassionate understanding of human nature as he was for medical skill. The persistence of his type was attested to by the good doctors in Sinclair Lewis's *Arrowsmith* (1925), Sidney Kingsley's *Men in White* (1933), and James Gould Cozzen's *The Last Adam* (1933). Dozens of practitioners in movies or soap operas on radio and on television have been in the tradition. Paul De Kruif's factual books about medical research, and two works indebted to them—*Arrowsmith* and Sidney Howard's *Yellow Jack* (1934)—showed doctors playing heroic roles.

In 1939, James T Thurber's *The Secret Life of Walter Mitty* hilariously burlesqued the clichés in such portrayals in a memorable passage. In it browbeaten little Walter fantasizes himself as a great surgeon performing a hazardous operation on a millionaire banker suffering from 'Obstreosis of the Ductal Tract, Tertiary' after the anesthetizer has busted and 'Coreopsis has set in.' The targets in this popular comic masterpiece of course aren't doctors but melodramatic and sappy portrayals of them, laymen's conceptions of their terminologies and procedures and—essentially—of Mitty's pitiful daydreams.

And of course literary comedians, noticing that at least some physicians weren't all bad, now and then portrayed them as *eirons* of the sort Americans admired.

That champion baiter of doctors, Mark Twain, for instance, in *Innocents Abroad*, handled over to a touring medic the role that he himself was most famous for assuming—'doing his best', as he put it, 'to conceal the fact that he even dimly suspects that there is anything funny' in what he's saying. Irritated by guides who wallowed in the admiration they excited by mouthing memorized speeches, Twain's mischievous tourists drive them wild by pretending to be stupid.

'The doctor', wrote Mark admiringly, 'asks the questions, generally, because he can keep his countenance, and look more like an inspired idiot, and throw more imbecility into the tone of his voice than any man that lives. It comes natural to him.' Then follows a generous sample of the physician's wonderful artistry—puzzled inquiries about a document in the handwriting of Christopher Columbus, about a bust of Columbus ('Ah—which is the bust and which is the pedestal?'), and finally some infinitely baffled questions about a mummy in the Vatican museum, ending: 'Mummy—mummy. How calm he is—how self-possessed. Is, ah—is he dead?'

In 1972, entrusted with editing comic gems published through the years by the most popular American magazine, *The Reader's Digest*, Clifton Fadiman told about his doctor's ironic joke: 'I learned a new word the other day while I was getting a medical checkup. The doctor told me I had idiopathic something-or-other. Alarmed, I asked him what "idiopathic" meant. He said, "Oh, it means we don't know a damn thing about the cause of the disease—indeed we don't know anything about it." This reassured me.' Quite a few of the *Digest's* favorite jokes showed doctors being humble about their capabilities, for example, 'Medicine', one physician tells a patient, 'has made great progress, Mrs Smedley. We used to think your trouble was caused by teeth. Now we've progressed to a point where we don't know what's causing it!'

A couple of recent comic television series have featured very popular doctors who are quite vocal about their shortcomings. In them both Danny Thomas and Bob Newhart constantly win laughs by making dry comments concerning their shortcomings—taking the pose of the *eiron* and letting on that they are less knowledgeable than in fact they are.

Of course, medical men themselves have known all too well from the time they started to practice that they aren't infallible: the news is that laymen seem to be giving them credit for knowing. An anthology of jokes from *Medical Economics* called '*It Only Hurts When I Laugh*', published a few years ago, has had wide circulation as a paperback. Its blurb announces that this is 'a rib-tickling collection [creating] feverish laughter and aching sides', and that its jokes were made by 'the nation's top specialists.' Predictably, 80 percent of the jokes that the M.D.'s tell are on themselves.

Despite these and other convincing proofs that doctors can be as self-deprecatory as anybody, in jokelore and satire the old joshing stereotype of the pretentious, boastful, and relatively ineffective doctor persists. Why? Well, for one thing, stereotypes never die or even fade away. There is a more revealing explanation: Though laymen—real and potential patients—aren't aware of the fact when they kid physicians in the age-old way, the ultimate joke is on them. Like most long-lasting japes, it has submerged elements of pathos and fear.

Humorists and satirists more often than not joke about worries or fears. Frontiersmen joked about hardships— huge and ferocious beasts, great distances, unclimbable mountains, horrible weather, ruthless Indians. When laymen, helped by psychologists, noticed that mental afflictions menaced everybody, humorists started to kid the newest publicized threat. A popular character came to be a Poor Little Man scared of everything. One humorist called his book *A Bed of Neuroses*; another dubbed himself and his colleagues 'the dementia praecox school.'

Nondoctors desperately want to believe that, even when a physician pretends or tries to be 'something more than he is', if he says he can cure any ailment he truly can. Shaw put it this way in his preface to *The Doctor's Dilemma*: 'Until there is a practical alternative to blind trust in the doctor, the truth about the doctor is so terrible that we dare not face it.'

## Thoughts about medical writing: IV. Can it be funny and medical?

[Editors of AACR shared the belief that medical writing is capable of vast improvement. This article is one of a series of communications from Charles G Roland, MD, former Senior Editor of *JAMA* to John T Martin, MD, Chairman Department of Biomedical Communications, Mayo Clinic, Rochester, MN.]

RBP from
*Anesthesia and Analgesia...
Current Researches*
50.2
March – April
1971.

Dear Tom:

It occurs to me that of all the emotional modes which one may bring to writing, humor must be the most difficult to conjure up at will. If I really knew how to do the trick myself, some evidence of that ability would have shown up before now in my writings. But even though I can't be funny myself, I do have a few thoughts about humor in medical writing.

Few human diseases really lend themselves to humor—humor publishable in a scientific book or journal, at any rate. And the author must bring special delicacy and tact to his subject, for misunderstanding is too easy and may produce offended sensibilities.

When is humor acceptable? I think the answer is, when the sick human being who is involved can be abstracted or hidden. For example, one author describes, in splendid but still amusing style, the reason why the sphincter ani must be preserved when performing surgery in that area. The account is classic.

> They say man has succeeded where the animals fail because of the clever use of his hands, yet when compared to the hands, the sphincter ani is far superior. If you place into your cupped hands a mixture of fluid, solid and gas and then through an opening at the bottom, try to let only the gas escape, you will fail. Yet the sphincter ani can do it. The sphincter apparently can differentiate between solid, fluid and gas. It apparently can tell whether its owner is alone or with someone, whether standing up or sitting down, whether its owner has his pants on or off. No other muscle in the body is such a protector of the dignity of man, yet so ready to come to his relief. A muscle like this is worth protecting [1].

Doubtless the anatomic site contributes to the humor here, as it does in the gentle pun used by one textbook author who reported that Enterobius is no respector of station in life, affecting even 'the seats of the mighty.'

Not only the seat, but also the feet can be the locus of a humorous description. In an article entitled 'Garlic-Clove Fibroma' the author reports a peculiar deformity of the toenails, and attempts to justify the descriptive name he has assigned to this disorder.

Numerous precedents exist for the use of a familiar descriptive term, a comparison with some household object. The following is a far from comprehensive list, exemplifying the graphic appeal of this type of classification: the silverfork deformity; the funnel breast; nutmeg liver; oat-cell carcinoma; scalloped ribs; cauliflower ear; port-wine stains; the doughnut pessary; sago spleen; current-jelly stools; the orange-peel dimple of breast carcinoma; the cherry spots of the ophthalmologists; strawberry tongue; coffee-ground vomitus; and bringing up the rear, the grapes of wrath of the proctologist [2].

And the last clause returns us to the ano-genital region, which certainly is the focus of most humor. I have already stated my feelings about the role of humor in medical writing, so I will close by quoting a colleague. Dick Reece summed up the case for humor when he said: 'I would never suggest that the physician writer not take his medicine seriously. I do hope, however, he will occasionally laugh at himself, amuse his readers when it helps to get and hold their attention, warm up his facts before he serves them, and use humor whenever it adds a touch of humanity' [3].

Regards,

Charles G Roland, M.D.

### References

[1] Bornemeier WC 1960 Sphincter protecting hemorrhoidectomy *Amer. J. Proctol.* **11** 48–52
[2] Steel HH 1965 Garlic-Clove Fibroma *JAMA* **191** 1082–3
[3] Reece RL 1967 Does Humor Have a Place in Scientific Writing? *Amer. Med. Writers Ass. Bull.* **17** 11–13

# Humor and the physician

*Fred D Cushner and Richard J Friedman*

RBP from the *Southern Medical Journal* **82.1** (1989) 51–52.

One might ask what the field of medicine has to do with humor. How can one find humor in a profession that deals with those who are sick, in pain, or dying? Although it may seem paradoxic, the combination of humor and medicine can be a worthwhile marriage for both the patient and the physician.

Upon closer examination, one can find that humor is deeply rooted into the medical profession. Henri de Mondeville, a famous medieval surgeon, proclaimed, 'Let the surgeon take care to regulate the whole regimen of the patient's life for joy and happiness. The surgeons must forbid anger, hatred, and sadness in the patient. Remind him that the body grows fat from joy and thin from sadness' [1].

Although many theories have evolved to explain the phenomenon of laughter, they can be summarized into three basic ones [1]. The *Derision Theory*, subscribed to by Plato and Aristotle, states that humor derives from man's readiness to laugh at the misfortunes of others. If this is true, physicians must have a good sense of humor, since they are daily faced with the misfortunes of others. The *Incongruity Theory* emphasizes the importance of sudden surpise, shock, and incongruity that trigger the laughter response. The *Liberation of Freedman Theory*, supported by Freud and Mindness, holds that laughter is a release from the constraints, fears, and pressures that all of us face daily. Therefore, fears of death and dying (or being hospitalized at a teaching institution in July) can be reduced via this release mechanism.

These days, it seems a good sense of humor is a prerequisite to becoming a physician. Without a good sense of humor, how could one spend 11 to 15 years in training, accumulate a $50,000 debt, and often work over 100 hours a week just for the privilege of being sued by the people one is trying to help? The fact remains that the job of a physician is both rewarding and stressful, and humor serves as an effective method to defuse those stresses inherent in the nature of the profession.

Reflecting on one's training, it is apparent that humor is deeply ingrained into the medical education process. The classical mnemonics learned during first year anatomy all have a comical twist, and those developed at the individual dissection tables are even more humorous.

189

Textbooks are ladened with humorous material. Interstitial emphysema has been described as air tracking into the subcutaneous tissue, causing the patient to swell into an alarming but usually harmless Michelin-tire-like appearance [2]. On the etiology of hepatitis, it has been estimated that the chance of acquiring the infection from oysters is one in 10,000; therefore, oyster lovers should not consume more than 9,999 in one sitting [2]. Discharge of flatus has been described as a symptom that can cause great psychosocial distress, and has been officially classified into three types: the slider, the gun sphincter, and the staccato [3]. While questions of air pollution and degradation of air quality have been raised, no adequate studies have been done.

It is not an exaggeration to say that Shem's *House of God* [4] is read by more medical students and house officers than *Harrison's Principles of Internal Medicine*. It is often quoted, with terms like 'gomer' and 'turf' well on the road to being included in standard medical dictionaries. The laws of the Fat Man are far better known than traditional laws such as Starling's.

Humor is quite common when house officers refer to specific rotations and hospitals to which they are assigned. A one-month rotation at the county hospital christens you as a 'County Mountie' while a rotation at the VA hospital entitles you to a membership at the 'Va-spa.'

No specialty is spared. Radiology becomes radiholiday, while nuclear medicine is known as unclear medicine. Gastroenterology is referred to as scoping for dollars and anesthesiology is known as doping for dollars. Urology probably has the most nicknames, such as stream-team and whizkids, to name a few. Orthopedic surgeons have to be strong as an ox and only twice as smart, and think that the sole function of the heart is to pump antibiotics to the bones.

Thus far, theories and examples of humor have been described, but what exactly is this phenomenon called laughter? It is a motor reflex, usually present by 4 months of age, that requires the coordinated movement of 15 facial muscles, as well as changes in the normal breathing pattern. The response is a predictable physiologic reaction, yet no neurogenic causes of inappropriate laughter have been described, such as MS, ALS, Alzheimer's, and pseudobulbar palsy.

As the smile begins, one sees a widening of the mouth as the corners pull up. The upper lip is then raised and the furrows from the external nares to the mouth edge curve down, rounding the cheeks. Creases develop under the eyes, and the eyes themselves develop a sparkling appearance, probably from reflex lacrimation, and there may be a slight vascular engorgement in the eyes due to an increase in the general blood circulation. This physiologic response is not confined to the face, for spasmodic muscle contractions, tachycardia, and increased catecholamine production also occur [5].

Recently, important questions have been raised concerning laughter in medicine [6]. Why do insurance companies not cover laughter if it is such good medicine? Since laughter is contagious, why hasn't the organism responsible been identified? Can a sense of humor be transplanted from one patient to another? Is it possible to laugh oneself to death?

In conclusion, humor and laughter are an integral part of both the life and the career of a physician. As stated by Mindness, 'Like love, courage, and understanding, humor is the one attribute that can sustain us through the worst.' By incorporating a style of humor into one's bedside manner, the stresses of illness and hospitalization can be temporarily relieved, and the overall doctor – patient relationship enhanced.

*References*

[1] Liechty RD: Humor and the surgeon. *Arch Surg* 122:519 – 522, 1987
[2] Robbins SL, Cotran RS, Kumar V: *Pathologic Basis of Disease.* Philadelphia, WB Saunders Co, 1984, pp 725, 900
[3] Berkow R: *The Merck Manual.* Rahway, NJ, Merck Sharp and Dohme Research Laboratories, 1982
[4] Shem S: *The House of God.* New York, Dell Publishing Co, 1981
[5] Black DW: Laughter. *JAMA* 252:2995 – 2998, 1984
[6] Buchwald A: On laughter. *JAMA* 252:3014, 1984

# Some hoaxes in medical history and literature

RBP from *Archives of Internal Medicine* **113.2** (1964) 291–6. © 1964 American Medical Association.

I really feel a bit self-conscious in writing about hoaxes in medicine. Some irreverent folk would say that medicine was a bundle of hoaxes and would point gleefully to our use of the placebo and to many of the trappings of the medical ritual. However, it is not that aspect of medicine that I want to discuss. That involves other considerations and an approach quite foreign to that in which I find myself at the moment. Then too I seem to detect in such banter a malicious undertone of cynicism, and it has been my experience that too much cynicism is in the end bad for the stomach. Rather, having due regard for the properties and in quite a lighthearted vein, I want to look at some of the hoaxes that are to be found in the medical records. My survey has turned up a few (my files contain many more outside the confines of medicine), but there must be a host of others, many of which will occur to our readers. I should add that this innocent little exercise in historiography was sparked by the receipt some months ago of a charming treatise on the subject [1].

Just what is a hoax? The dictionary defines it as a humorous or mischievous deception especially in the form of a fabricated story, adding that the word is probably a contracted form of the word *hocus*. It must be carefully distinguished from a fraud or an act of deception. The difference lies essentially in the purpose and the effect involved. A first-class hoax is carried out for amusement, in the spirit of pure mischief. Other more elaborate forms of deception such as a swindle, a fraud, or quackery have as their motive gain, power, or even revenge. The hoax is in essence a practical joke. Skilfully conceived and executed, it may be a work of art.

Of course the hoax is made possible by the wide prevalence of credulity in man, a trait which has been exploited since the beginning of time. There is present in all of us the will to believe, and, so often, to believe not wisely but too well. This weakness has provided a fertile province in medicine where charlatanism has flourished and given rise to some of the most bizarre chapters of the human comedy. But that in passing only; such reflections are outside our present province.

Hoaxes have been perpetrated in every sphere of life, and many of them provide the light relief of history. Lucian of Samosata, the Graeco-Roman satirist, in the form of a letter to his friend Celsus, in the year 180 AD, tells of the exploits of one, Alexander, from a town in Asia Minor, who by trickery was able to produce a snake from a goose egg, maneuver it across the threshold of the temple, and thus proclaim to the assembled crowd that this symbol of Aesculapius was a sign of grace from the gods of healing. In the early 18th century students at the University of Wurzburg hid clay fossils in the digging areas of their professor of zoology, Professor Beringer, who swallowed the bait, published a book on his revolutionary discoveries, and never recovered from the consequences when the students confessed. In literary annals there is the case of James Macpherson who in 1760 fabricated some originals and liberally edited others, and published work as translations from the Gaelic of Ossian. This act aroused the ire of the formidable Doctor Johnson and called forth some of his finest invective.

A really trenchant hoax may be recalled in the episode of Swift's Bickerstaff papers. In London a man called John Partridge, who claimed to be an astrologer, was publishing predictions in the form of almanacs, which prognostications annoyed Dean Swift because they, he felt, were simply Whig propaganda. Accordingly in 1708 Swift produced a rival almanac in which he forecast the death of Partridge on March 29. Then on March 30 he published an eyewitness account of Partridge's death together with a funeral elegy. Partridge indignantly protested and advertized in the papers that he was still alive. Swift retorted in a 'Vindication' proving that the poor man was really dead, and other writers took up the cry. In the end no one believed the unhappy Partridge, and so in effect he ceased to exist.

In the annals of elaborate hoaxes which later grew into a large enterprize there might, I suppose, be cited the stories of Leo Taxil who started by reporting the presence of sharks in the harbor of Marseilles and went on to fool many of the dignitaries of Europe, reaching the height of his notoriety in 1887; and of Madame Blavatsky (born as Helena Petrovna in the Ukraine) who successfully pyramided her skill in fooling the credulous and in dealing in Paracelsian occultism into the cult of theosophy which is still flourishing. But these are a far cry from straightforward good-natured hoaxes of the kind which we have in mind.

A prime example of this sort is the glorious hoax which was pulled off against the British government when in 1910 a party purporting to be the Emperor of Abyssinia and his suite reviewed the British Channel Fleet then lying off Weymouth, the ceremony being carried out with full pomp and circumstance. When it was made known that the visiting celebrities were in fact a group of English young folk, headed by the prince of modern English jokers, Horace de Vere Cole, most embarrassing questions were asked in Parliament, and the whole country laughed. It is also worth noting that one of the conspirators was a girl, Virginia Stephen, whom the world came to know later as Virginia Woolf. Incidentally, it was this same master of the hoax, Cole, who on another occasion scattered horse droppings about the Piazza di San Marco in Venice, that horseless city, and then sat back in a café chair to enjoy the spectacle of puzzled Venetians looking at the sky, then at the pavement, and then giving up in despair [2].

My memory recalls two hoaxes which figured during the early years of the first Great War. The first is the legend of 'the angles of Mons' which were said to have appeared during the battle of Mons, a legend which has since been shown to have arisen as the result of a short fictional story which was published about that time. The second is the tale told with bated breath of Russian soldiers who, early in 1915, had been seen in thousands passing through London on their way to the Western front. I myself remember being given this heartening news on my arrival in England in company with Canadian troops. Doubtless the whole yarn had its origin in the brain of some aspiring newspaperman, but it proved to be remarkably persistent.

The late James Bone, the beloved London correspondent of the *Manchester Guardian*, was a master of the classic hoax, and I have been regaled on many occasions by my newspaper friends reciting his exploits. On one of his rounds he noted a cat sitting on the counter of a little branch post office in Cheapside; whereupon he wrote a spectacular article drawing the attention of the London burghers to the fact that His Majesty's Government had provided a stamp-licking cat in this office to assist customers in dispatching their letters. The little office was swamped with customers for weeks afterwards. On another occasion while going about the National Gallery in London he noted a particularly dull picture portraying the Madonna and Child. He went home and wrote a story which appeared the next day stating that a miracle was daily being enacted in the National Gallery, in which, mentioning this particular picture, at a certain hour of the day, tears were observed to roll down the cheeks of the Madonna. The result—the unheard of spectacle of queues outside the Gallery and a healthy increase in the attendance at that venerable institution.

It was still another hoax that provided the genesis of the esoteric organization known as The Baker Street Irregulars, the Sherlock Holmes Society devoted to the study of the master detective, which now has Chapters in many centers on this continent, and the allied organization, The Sherlock Holmes Society of Great Britain. In 1928 Father Ronald A Knox, the Oxford scholar, published a book of essays one of which was entitled 'Studies in the Literature of Sherlock Holmes' [3]. This was a tongue-in-cheek satire of the Higher Criticism books and articles then much in favor in Europe's scholastic circles. The idea was taken up by Christopher Morley who at the time was conducting the Bowling Green department of the *Saturday Review of Literature*; presently a society was formed whose purpose was the study of the Sacred Writings recorded by that famous medical man, John H

Watson, MD, who consummately played the Boswell to his companion, Sherlock Holmes. The society soon attracted the Sherlockian wits of the day, scion groups were established, a journal was launched, and the Fellowship has grown to this day.

It is against this colored backdrop that I should like to mention a few modest medical hoaxes. What launched this farrago in a serious medical journal is something of a mysterious story. Following the publication in these pages of an article entitled 'William Osler: Obstetrician' (May, 1962) in which I discussed Osler's famous Baby-on-the-track case, I received at intervals of about a fortnight three off-prints of articles by one S N Gano, which had appeared in *The Leech*, a journal published by the Students' Club of the Welsh National School of Medicine at Cardiff. In each case the item was sent from a different city in the United States, and no indication was given of the forwarding agent. Any wonder that I may have had at this feature was soon forgotten in the delight that the essays provided, two of which were beautiful 'spoofs' of heavy scholarly scientific 'communications'.

Let us look at the first of these learned essays. It is titled 'A Gloss Attributed to the Hippocratic School' [4], and purports to be a short Hippocratic text which had been overlooked by such scholars as Littré, Jones, Adams, and Withington. Dr Gano states that the original is thought to have been at one time in the library of the Church of Santa Eufrasia in Pisa, and goes on to remark that 'the stylistic traits bear strong evidence of genuineness, and the reader cannot fail to be impressed by the conciseness, clarity, and sagacity which are displayed in almost every paragraph.' The essay is equipped with the full apparatus of scholarship with cunningly spiced footnotes. The editor of *The Leech* in his introductory note explains that the author, S N Gano, is the journal's 'special correspondent in New York'. I have been unable to obtain any further information from the office of the journal at the medical school in Cardiff. So that it seems that a satirist of rare quality is lurking in anonymity either in Wales or, less probably, in

196

New York.

The text of this rare incunabulum deals with: Prognostics, Epidemics, Aphorisms, In the Surgery, Fractures and Articulations, and Ancient Medicine. Each observation is numbered. Space does not allow me to quote the whole, but here are some excerpts. The voice is the voice of Hippocrates, but the hands are the hands of the great contemporary scholar, Gano.

1. Absence of respiration is a bad sign.

2. It is unfavorable for the patient to be purple, especially if he is also cold. The physician should not promise a cure in such cases.

4. Hemorrhoids are not improved by horseback riding or by riding on an ass.

7. Drowsiness and the itch are incompatible.

10. On the island of Tenebros, in the spring of the year, several maidens were attacked by swelling of the abdomen and absence of the menstrual discharges. Giddiness; face flushed; absence of the membrane. A voyage to Asia was recommended. Complete crisis at the end of the ninth month.

11. Patients who always wake up on the wrong side should be treated by purges.

13. To pick at the coverlet indicates alienation of the mind, unless the bed contains small fragments of food.

15. Fevers of eighty days where no cause is apparent produce alienation of mind in the physician.

22. Life is short and the art long; patients are inscrutable; their ignorance is impenetrable, and their relatives are impossible.

23. When swarthy patients have a blonde child and a blonde serving-maid, the physician should suspect a displacement of humors. I hold that this condition is no more divine than any other.

25. When the spleen is found on the right side, the patient should consider changing physicians.

30. Dislocation of the neck, which has been produced by a rope, cannot be treated with barley water.

40. Where symptoms are severe and protracted, it is good if the patient's relatives are few, and best if they be absent altogether.

The second essay reprinted from *The Leech* (Autumn, 1960) bears the title 'Deficiencies in the English Medical Vocabulary'. Dr Gano's thesis here is that in clinical practice our terms are inadequate to satisfy every need. He gives several instances. The first: 'Let us suppose that the clinical clerk or house officer on his ward-round finds a patient who has protruding eyes and protruding ears. What shall he set down in his record? The patient has exophthalmos and also ...? *Exotosis?* No, this would lead to confusion with *exostosis*, and hence with boneheads.' Surely, he argues, a technical term is lacking here. Then again he points our that we are deficient in necessary verbs. Instead of asking whether the patient is dyspneic, would it not be much more concise to ask, 'Does she *dysp?*', when the reply would be, 'She both *dysps* and *orthops.*'

The absence of intensive verbs is especially acute in the sphere of circulatory disorders. In the case of the physician whose patient is developing coronary thrombosis, should the doctor not be able to have words at his command without being reduced to saying, 'He is *thrombosing*', or 'He must be *anticoagulated*'. And later, if the kidneys fail, 'He must be *dialyzed*'. Similarly the physician may understandably say of the patient: 'He is fibrillating' or 'fluttering'. But suppose that the patient only has ventricular extrasystoles? Can he say that his patient is *extrasystolizing*, or, if he is really to be scientific, *ventricular-extrasystolizing*?

In other conditions we clinicians are similarly handicapped. Why clumsily have to ask if the patient ever suffered from icterus when the more direct way would be to inquire, 'Did she *ict?*' Or under other circumstances, 'Does she *dyspareuniate?*'

After much else in this exercise of looking at medical practices with a fresh eye, our critic ends with these words:

198

And what of that paragon and prototype patient, the 'Well-developed and nourished white male'? The progress of literature now demands a newer term, and the progress of science requires that vagueness be corrected by precision. There are so many kinds of well d. and well n. wh. males. The species should be broken up into its principal sub-species, thus: an *Apollo*, an *Achilles*, a *Thersites*, and, of course, a *Bacchus*. Let art and science join hands and march forward.

The third item is equally diverting and, as I have been told, had a most remarkable sequel [5]. The author in this case does not appear to be the scholarly Dr Gano, but I suspect one of his colleagues. (Unhappily, as I have said, documentary evidence is lacking.) This is a biography of a certain Dr Emile Coudé, the alleged inventor of the *coudé* catheter, adorned with the customary footnotes and a photograph. The whole is obviously quite spurious. But the sequel is delicious. At the time that the paper appeared Sir Hamilton Bailey was preparing a new edition of his book *Short Practice of Surgery*. He incorporated some of the Coudé biography into his text, and it was not until the book was in page proof that the deception was detected. In consequence several pages of the new edition had to be reset. The incident was suitably celebrated by several letters in the *Lancet*, thus placing the laurel on a successful practical joke.

I am indebted to Dr Saul Jarcho, whose article I have already referred to, for directing my attention to some other notable incidents which are in the genre of the hoax. Sir William Osler's propensity for practical jokes has become legendary, and sometimes it got him into trouble. Harvey Cushing's *Life* of Osler may be consulted for the details. Osler used the name of his alter ego, Egerton Y Davis, as the cover for his pranks. One of the items which he published under the name of the versatile Dr Davis takes the form of a letter to the editor of a medical journal, *Medical News*, in 1884, describing a case of vaginal spasm, with what my correspondent terms 'Rabelaisian' complications. Osler's colleague, William Henry Welsh, was also a hearty exponent of the art of the practical joke which seems to have flourished in the '80s and '90s of the last century. He frequently pretended to a sympathetic physician that he suffered from aortic aneurysm, and it is recorded that on one occasion he concealed a rubber bulb in his clothing in order to alarm his friend, Halstead, who was examining him.

I can vouch for the truth of an incident which took place in the first decade of this century on a train between Toronto and Montreal. My aunt was travelling on the train in the company of a physician who, being of a lively and imaginative disposition, became bored with the gloom of the journey and his impassive Canadian fellow passengers. In order to enliven the situation he simulated an epileptic seizure with terrifying artistry, leading to tremendous confusion among passengers and crew, to my poor aunt's discomfiture, and to his own evident glee and satisfaction.

Encyclopedias and dictionaries of biography have always been a tempting area for hoaxers. There are many instances of entries in which artists in make-believe really let themselves go. In the medical world Appleton's *Cyclopaedia of American Biography* is a case in point. This work was issued in seven volumes between 1887 and 1900. Subsequent investigation has shown that it contained at least eighty-four biographies of celebrities who never existed, including a man who was given the palm of fame by virtue of controlling a cholera epidemic which a checking of sources was shown never to have occurred. Some one or more of the compilers could not resist the temptation to thumb their nose at history.

Another hoax the echo of which is still with us is the handiwork of the late H L Mencken [6]. To divert himself he printed an article in the *New York Evening Mail* for December 28, 1917, carrying the title 'A Neglected Anniversary'. He stated that seventy-five years before, a man, Adam Thompson, had installed in his home in

Cincinnati the first bathtub in America, and added the other historical item that the first bathtub in the White House was set up in 1851 under the order of President Millard Fillmore. He then went on to describe the storms of protest from citizens, medical societies, and state legislatures, and the reluctance which was shown in finally adopting the new device. The story was believed and became the staple of newspaper editorials, standard reference works, medical articles, and moralizing sermons. Nine years later in 1926 Mencken, having, as he says, 'undergone a spiritual rebirth and put off sin', made public confession of his hoax which was printed in thirty papers. But all to no avail. Some time later he printed a second confession, but with little effect, for the 'tale of the tub' goes on to this day. At least two reflections emerge from this story. The first: Here is one way in which legends are born (King Alfred burning the cakes). And the second: The moral, my dear Horatio, is to check (and double-check) your references.

These have been variations on a bizarre theme. There are, of course, many other comments which might be made. It would seem that the hoax is the product of a more innocent age than our own sophisticated era. I suspect that many hoaxes are embedded in our medical books, in footnotes particularly, and in vulgar errors which have been solemnly copied out from generation to generation of medical commentators. Some of these errors may still be around today. It follows too from our recital that the really good hoax is a work of art, and in certain circumstances deserves a place in social history, if only as a satire on existing customs and institutions.

Finally, a hoax now and then may be relished by the wisest men. But the hoax soon palls, and too frequent practical jokes are intolerable and must be put down by every means short of homicide. Under such provocation the victim may be forgiven if like the Host, after hearing Sir Thopas's ballad recited over and over, he exclaims—'No more of this, for Goddes dignitie'. And the same thing applies to articles dealing with the hoax. Wherefore I bow out abruptly without further ceremony.

E P SCARLETT, MB
Calgary Associate Clinic
Calgary, Alberta, Canada

References
[1] Jarcho S: Some Hoaxes in the Medical Literature *Bull Hist Med* 33: 342–347, 1959.
[2] Highet G: The Art of the Hoax *Horizon* 3: 66–72, 1961.
[3] Knox R A: *Essays in Satire* (London: Sheed and Ward) 1928.
[4] Gano S N: A Gloss Attributed to the Hippocratic School, *Leech* 4: 17–19, 1958.
[5] R P: Emile Coudé (1800–1870), *Leech* 6: 15–16, 1957.
[6] Evans B: *The Natural History of Nonsense* (New York: Alfred A Knopf) 1946.

# Fun at the expense of specialists

Surgeons do it.
Internists talk about it.
Radiologists just look at the pictures.

Principles of dermatology:
If it's wet, dry it.
If it's dry, wet it.
If neither of these works, use steroids.
If steroids don't work, do a biopsy.

Patients come to ophthalmologists about their eyes but also for other things. This woman came into my office and looked around to make sure no one was listening. She said, 'I really am here about my eyes, but I have another problem maybe you could help with. I have terrible trouble holding my water. What do you advise?'
 'I advise you to get off my carpet.'

Military medicine bears the same relationship to medicine as military intelligence does to intelligence.

Q: What's the guiding principle of orthopedic surgery?
A: 'Ugh, me cut.'

Two psychiatrists meet on the street. One says to the other, 'You're fine. How am I?'

Q: How many psychiatrists does it take to change a light bulb?
A: Only one, but the light bulb really has to want to change.

*Surgeon*: You need an operation.
*Patient*: I think I would like a second opinion.
*Surgeon*: OK. You don't need an operation.

I DO NOT UNDERSTAND QUANTUM MECHANICS. BUT THEN I'VE <u>ALWAYS</u> HAD DIFFICULTY WITH THEOLOGICAL ARGUMENTS.

# Conferences, Reports

The most prominent requisite to a lecturer, though perhaps not really the most important, is a good delivery; for though to all true philosophers science and nature will have charms innumerable in every dress, yet I am sorry to say that the generality of mankind cannot accompany us one short hour unless the path is strewed with flowers.

*Michael Faraday*
Advice to Lecturers

# How to act at a seminar (from the cycle 'Useful advice to scientists')

[The following two pieces were selected and forwarded by Academician and Nobel Prize Winner N G Basov. They have been translated by Professor Lorraine Kapitanoff.]

RBP from Yu V Afanasiev, illustrated by V V Mikhailin, *Priroda* No 6 (1981) pp 126–8.

We are always glad to help you and to give valuable advice. But first of all we would like to note what a seminar is:

1) it is not a theater
2) it is not a club of jolly and resourceful fellows
3) it is not a circus

but a meeting of critically attuned scientists who are still not rested up from Saturday and Sunday (if the seminar is on a Monday) and are still entirely fresh (if the seminar is on a Friday) among whom are: your chief (first rank); your rivals (second rank); several of your good buddies (last rank) and everyone else who thinks that a seminar is a theater, a club of jolly and resourceful fellows, etc.

Not ruled out also is the presence at a seminar of the well-known wag and jester who brings into the discussion elements of freedom and miracle-working. Therefore it would be better for you not to go to the seminar. The ancients already wrote:

*Ut stultitia etsi adepta est quod concupivit nunquam se tamen satis consecutam putat sic sapientia semper eo contenta est quod adest, neque eam unquam sui poenitet.†*

But if it so happens that you must attend a seminar then without fail follow our 'Useful advice.'

## I. If you are a young, energetic investigator...

You arrive two hours before the beginning and cover the entire blackboard with formulas (if you do not have enough of your own formulas, copy from any handbook) but in no case use them during the lecture. It is entirely sufficient, having made a vague gesture in the direction of

---

†In any case let us draw a conclusion. And if stupidity even having attained what it hungered after, nevertheless never considers what it acquired sufficient, then wisdom is always satisfied with that which exists and never is vexed at itself.

the blackboard, to say 'We here have already calculated...'

Your outer appearance is very important. Already the elder Du Belle said 'To me the scholarly appearance of a pedant is abhorrent.'

Therefore it is necessary to be dressed in jeans and an old sweater. A soiled handkerchief and sneakers without laces worn on bare feet doesn't look bad either nor does a not entirely fresh striped vest.

The presentation must be begun with a simple, artless (but not flippant) smile by which you demonstrate your recognition of your insignificance in the presence of Great Scientists. Your talk must be brief, as indistinct as possible and, primarily, unintelligible. The situation in which you yourself do not understand what you are saying is ideal. This appears very natural and unaffected. It is very useful to make a gross error in an elementary formula. It's not ruled out that someone from the first rank will notice it and point it out to you with satisfaction. In any case it is forbidden to distinguish the primary results from the secondary ones and from the ones already well known to everyone—leave this to the listeners. To observe this rule is especially important in the absence of results. During an explanation of points which you don't understand at all, it is necessary to switch from words to gestures, i.e. with the help of chaotic movements of the arms and legs (and, with appropriate training, also of the ears), shuddering, increased breathing and opening of eyes wide, and rolling them, reversing of the head and sticking out your chin to transmit to the listeners your unshakable faith in the omnipotence of science.

It does no harm to drop the chalk several times, to step on it, to dash about the hall in search of a pointer, not to light the light after showing the next slide or on the other hand to show slides without extinguishing the lights, etc. However, the report is still not the main thing. The main thing is the questions. It is always necessary to answer them. Our experience shows that such and only such cases are possible.

**1st** You understand the question and know the answer (this is a joke; such a thing never happens).

**2nd** You understand the question but do not know the answer (extremely rare). Then you begin: 'Already in 1965 someone (your chief) showed... (go as far as you like). They will immediately interrupt you and in the process of the discussion everything will rapidly get out of hand and they will generally forget about you.

**3rd** You do not understand the question (normal situation). Here you make a wise face (if this is possible), with an open, firm and straightforward glance look at your questioner and say 'In the present study we have undertaken an attempt only to show that... (repeat word for word the colophon as many times as is necessary). If the questioner is not calmed down then your leader will answer the question.

**4th** A question is posed by one of your friends (you naturally have a prepared answer).

## II. If you are a well-known scientist, fatigued by overseas service

First of all you must wear a suede or velvet jacket. A pipe produces a strong impression. (With special training it is possible to learn how to deliver the entire lecture without taking the pipe out of your mouth.) The hands (or at least one of them) must be kept in the pockets. Going to the blackboard, it's not bad to exchange some joke or other with friends. After approximately one minute you, with a melancholy smile, look out the window and say wearily, 'The first slide, please.' For still another minute look absently at it, and finally intone: 'This is not at all

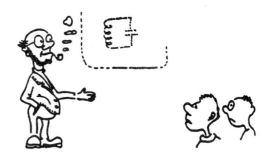

what you have, no doubt, already surmised. This is:

1) a self-congruent degenerated system of equations (of quantum electronics, gas dynamics, dynamics of lasers) in an entirely divergent form, or

2) a block-diagram of a new super multi-channel laser, thermonuclear device with a commutating power recording apparatus.

After this, immediately ask that the second (and last) slide be shown, on which there must be a composite graph in arbitrary, relative units. In any case it is forbidden to state on which axis it is plotted. At this point it is necessary to digress and to tell of the impression which this slide made at the meeting of the American Physical Society. It is useful to briefly illuminate further the history of the problem and in an easy, ironic manner to emphasize all the fruitlessness of the attempts of other authors, to even understand the essence of the problem. In conclusion it is possible to concentrate attention on some sort of trivial question unrelated to the business at hand (which you are well aware of) and to write 'expromptu' an approximate formula. At the very end it is necessary to become thoughtful for a couple of minutes (during this you can light up your pipe).

To answer questions (irrespective of their content) one must act approximately as follows:

**1st** (sternly) 'This automatically follows from the equations.'

**2nd** (edifyingly) 'Let's leave off these trivial contrivances.'

**3rd** (condescendingly) 'We calculate this effect effectively.'

**4th** (ironically) 'I somehow don't understand your qualitative reasoning.'

---

## Conference in conception

[Much has been written on how to organize a successful scientific meeting, so much so that the unsuccessful conference remains largely unexplored. The author here attempts to redress the balance with some personal guidelines.]

RBP from *New Scientist* 20/27 December 1979 p 954 (condensed).

1. The *name* of the meeting is to be chosen with especial care. It must incorporate a search for a sense of cosmic awareness with a profound feeling of aesthetic congruence.
2. The *date, place,* and *duration* of the meeting need very careful thought. Few so far have been courageous enough to meet for three weeks in August in a tented camp on Kerguelen.
3. The *Registration Fee* is of singular importance. It should bear a close relationship to the numbers the organizers propose to attract. While it is reasonable to work on a clear profit of $100,000 for the larger meetings, it is usually practicable to make $20,000 on the smaller.
4. *Abstracts* should be demanded at least a year in advance. This will ensure that they are written before the papers.
5. *Language.* It requires both courage and enterprise to insist on the use only of, say, English and Basque; but to do this for a meeting in Togoland verges on the eccentric.
6. *On arrival,* in return for the registration fee, the attending member is to be presented, preferably by a young woman of outstanding charms, with a plastic container not quite large enough to hold its ponderous contents of 20 documents, 18 of which should be colourful but inaccurate descriptions of local 'beauty

spots.' A list of attending members should on no account be included.

7. Deeply hidden among the container contents there should be a *label*, preferably misspelt. It is usual for some *device of attachment* to be incorporated in the construction of the label. Though there is no objection to this being of a nature likely to damage either skin or clothing, it is normal to aim at only one.

8. Ideally *concurrent sessions* should be held. Preferably located in totally separate buildings some distance apart, these may otherwise be located at two ends of a labyrinth. It is essential that the failure of one speaker to turn up in time is *not* followed by a gap until the next one is due, but immediately by him, if he is there; and if not *by any other speaker.*

9. The *time permitted for individual presentations* is to be calculated using the Blenkinsop – Throssingfleet formula: by dividing the number of conference-days minus three, multiplied by 17, by the number of lecturing participants, itself divided by the number of concurrent sessions plus $\pi/k^2$, the constant $k$ being 4.732 for meetings north of 42°N and 3.778 for those elsewhere.

10. *Recalcitrant lecturers who persist in attempting to go beyond the time limit* need brusque treatment. Usually effective is temporary acoustical stunning, either orally or by a mechanical device. Bright lights may also be shone into the eyes.

11. All projection of *slides* is best done by hired workers with defective eyesight. This avoids emotional involvement.

12. In view of past confusion over *slide marking*, it should be insisted upon that all slides be marked with a large red circular spot one centimeter in diameter, exactly centrally, and that they are of identical appearance whether viewed from the back or the front. The minor inconvenience this occasions is more than balanced by the supreme confidence it gives to both projectionist and lecturer.

13. All *diagrams for overhead projection* should be drawn on thick, good quality tracing paper. In case this does not entirely prevent emergence of an image, a vase of gladioli strategically placed is a valuable investment and a trifling expense.

14. It is a gross extravagance to provide a *projection screen* of large dimensions. One a metre square, or a trifle larger, is quite adequate. A larger display is apt to be dazzling. Attempts by a minority in the back rows to use field-glasses should be foiled.

15. *Discussion* should, as far as possible, at all times be prevented.

16. It is usual to *inaugurate* the meeting by a session to which are invited at least 20 members of the local or national leaders of the community. These should be as distinguished for their total irrelevance to the subject as they are for an ability to pontificate, and should have a solemnity of appearance as impressive as it is genuine.

17. Early in the proceedings an *evening reception* is usually held. It is always possible to charge extra for this, and the locale needs to be chosen carefully so that the majority of those attending arrive after consumption of the choicest wines and comestibles.

18. A *banquet* later in the proceedings gives a wonderful opportunity to add to the profits, even if they have to be shared with the caterer. A special table for the organizing committee is of course essential.

19. Any *field trip* should be phenomenally expensive. It might lead to discoveries which could seriously damage existing dogma.

20. Though it has been common practice for some years to have a concurrent *Ladies meeting*, with excursions and perhaps also a *Children's meeting*, few organizations appear to have recognized the potentialities offered by one for the pets of attending members. Unfortunate is the dog which has not had the emotional experience of the sight of the Place de la Concorde, with its 472 lamp posts.

*Note:* Problems and opportunities of *allocation of accomodations* and *group air travel arrangements* will be discussed in a later paper.

---

## If summary papers incline to prosaic then try dactylic or even trochaic

*Science* **216**: 167, 9 April 1982, © AAAS. Reproduced by permission.

The end of the conference approaches and the inevitable early trickle to the airport turns into a flood. A much-diminished and more than a little distracted audience hears out the summary speaker, more from politeness than in expectation of actually learning anything, still less of being entertained. Sounds familiar? Of course it does.

So, when Donald Patterson, of the University of Pennsylvania, was asked to summarize an International Symposium on Animal Models of Inherited Metabolic Disease,† he decided he would break the pattern—he delivered his summary entirely in verse.

Papers at the meeting ranged from 'H-Y antigen and intersexuality in animals', through 'Mucopolysaccharidoses', to 'Some animal models of lysosomal storage diseases.' Weighty prose for distillation into poetry, it has to be admitted. 'It's really just verse rather than poetry', suggests Patterson. 'I write verse for a hobby, mainly to do with medical things.'

Only a few close colleages knew of Patterson's plans. The organizers of the National Institutes of Health meeting were unaware that their conference would take a lyrical turn. Patterson worked on his preparation before the meeting, using submitted abstracts. 'I couldn't have done it all during the meeting', he says, 'so I had to try to guess what people might say in their papers. I did put some finishing touches here and there, mainly in the

†The meeting, organized by the Registry of Comparative Pathology of the Armed Forces Institute of Pathology and the Universities Associated for Research and Education in Pathology, Inc., was held at the National Institutes of Health, 19 and 20 October 1981. Poems reprinted from *Institute of Laboratory Animal Resources News*, **25** (No. 2), 6 (winter 1982).

evenings but sometimes during the papers themselves.' The result of these labors was more than 20 poems with as crisp a distillation of difficult science as could be wished for (of which, some examples follow).

Patterson's favorite? 'I like the one on "Histocompatibility, disease, and aging",' he says. 'I don't know why; it just seems to go nicely.' Alas, the paper's author, Edmond Yunis, of the Sidney Farber Cancer Institute, Boston, had left before summary time came around.

In case conference organizers nationwide should already be planning to invite the Pennsylvania poet to add some style to forthcoming gatherings, Patterson points out that this was his 'first—and last—venture into the poetic summary.'

## Gene structure, organization and expression

*A W Nienhuis*

*Well, we thought that we finally had figured it out!*
*We knew what the structure of genes was about.*
*'It's simply a piece of the old DNA,*
*Transcribed and translated the usual way;*
*The way that Jacób and Monód always said.*
*It's simple,' we said, 'when it gets through your*
*   head.'*
*But there's two kinds of Karyotes—there's Eu- and there's*
*   Pro-*
*And what's true for E. coli just isn't so,*
*When it comes to the genes of a mouse or a man.*
*Mother Nature, it seems, has used more than one plan.*
*Dr. Nienhuis informs us it's much more exotic*
*When dealing with animals Eukaryotic.*
*Their chromosomes aren't just ribbons of genes,*
*One coming right after the other, it seems.*
*Eu-genes are in pieces—they're really quite split.*
*A Eu-gene's got introns in the middle of it.*
*And this complication just leads us to more:*

214

*RNA now needs cutting and splicing before*
*It can serve as a template, appropriate to*
*The making of proteins. That's one* thing we knew.

## Hemoglobinopathies—from phenotype to genotype

*W F Anderson*

*We'd like to explain what pathology means,*
*In terms of what's wrong with the structure of genes;*
*Know if a control or a structural locus*
*Constitutes the exact pathological focus.*
*Dr. Anderson's talk has made it quite clear*
*That the answers to some of these question are near.*
*At least with respect to the globins, we know*
*Why some mutant's erythrocyte levels are low.*
*With the help of the enzymes that slice DNA,*
*And cloning techniques, we now have a way*
*To study the actual sequence of bases;*
*To know when those purines are not in their places.*
*In humans who have a resistant anemia*
*That goes by the general name, thalassemia,*
*Globin genes can be missing, we don't know where they*
    *went—*
*Perhaps an unequal crossover event*
*Has caused their deletion—whatever—they're gone.*
*In others they're present, but never turned on.*
*The latter are viewed with much more expectation*
*As keys to the problem of gene regulation.*
*What's needed are animal models of these*
*So look, animal hematologists, please!*

## Histocompatibility, disease and aging

*E Yunis*

*'The crown of life, our play's last act,'*
*Cicero on old age was opining.*

*What he didn't know, but now is a fact:*
*It's then your T-cells are declining,*
*Too many tick-tocks of the old thymic clock;*
*It runs down like a watch on the shelf.*
*Then suppressor T-cells aren't sufficient to block*
*B-cell clones that arise against self.*
*This theory's supported, Dr. Yunis explained,*
*By studies in mice and in man.*
*The data suggest that the program's ingrained;*
*It's a genetic kind of a plan.*
*It seems to depend on your HLA type.*
*If you have a desire to die late,*
*And your wish is, in time, to become overripe,*
*It is better not to B-8.*

# Modern type concepts in entomology

*N T Baker and R M Timm*

*J. New York Entomological Soc.* **84**, September 1976, No 3 p 201. RBP of The New York Entomological Society. Condensed.

**Abstract.** Twenty-six new type concepts are proposed to alleviate the barrenness of current type methodology. The proposed type concepts are commonly used and practiced but completely unrecognized in the scientific literature. The truth inherent in these new proposals will be patently evident and should be given due consideration in light of current systematic procedures.

The type concept serves as a standard of reference to tie taxonomic names to objectively recognizable taxa. The standards are the types, and the types are only the specimens bearing the name of the taxon. A type is always nothing more than a zoological object. In this regard and for the sake of standardization in systematics in general, only three type concepts (holotype, lectotype, and neotype) currently are exercised in the type method. This standardization unfortunately has a certain sterility. Additional type concepts could alleviate such a situation and simultaneously do systematic science a real service.

These proposed type concepts and their use were hypothesized, distilled, crystallized, and recrystallized through creative deliberate debate with members of the Association of Minnesota Entomologists and several other distinguished entomologists. They were found to apply in one situation or another where no other concept seemed quite appropriate. With few exceptions, *all* these type concepts have been found to have a usefulness apparently unappreciated by the systematic entomologist. The reader should be aware that these proposals are an attempt at satirical humor on taxonomic entomology.

*Ambiguotype.* 1. A type specimen, usually a holotype, with inadequate date-locality labels. Classics are: 'N. Amer.', 'Northwest Territory', 'my backyard', 'Summer 69', 'Highway 313', etc. 2. Also known in some circles as a type based on a 'Walker description' or as a 'Walker type'.

*Artotype.* Type specimen of a new species with distinctive color patterns ultimately shown to be paint spots.

*Autotype.* Holotype collected from the grill or radiator of your car.

*Boobootype.* A holotype that should not have been described. In this case, a specialist fails to recognize his own earlier described species; the museum technician or the star graduate student does, however.

*Chromosomotype.* A type specimen of a new species known only from its chromosome smear because the remainder of the specimen was discarded.

*Dermestotype.* A holotype usually consisting of only a partial thorax and some attached legs topped by artifacts reminiscent of a dermestid orgy.

*Diplomatotype.* A type named for someone with whom the describer wishes to have a good rapport, i.e. *Nationalsciencefoundationulus, NIHulus, Racquelwelchae*, etc.

*Dissectotype.* The type specimen of a species recognized to be new after you have dissected the beast entirely. Often the dissectotype can be cleverly converted to a Dermestotype.

*Incognitotype.* 1. The type specimen which is positively the holotype but has lost its identifying labels. 2. A type created when the holotype is deposited in a personal collection which ultimately disappears. 3. A holotype, presumed lost, which is needed for a major taxonomic revision.

*Patronymotype.* A holotype in the personal collection of a collector who will relinquish it to a recognized authority if the new species is named after the collector and he gets to keep the holotype.

*Pornotype.* A type category frequently used in entomological circles (but never recognized in the scientific literature) in which systematic decisions are based predominately on extensive examination of genitalia. In some cases, it is difficult to ascertain whether pornotype is a matter of personal taste or a necessary professional evil.

*Tyrannotype.* Type designated by the International Commission on Zoological Nomenclature. Apparently a necessary evil.

---

# Liposome letters *et al*

[This is the Preface to *Liposome Letters* by the editor A D Bangham, 1983, as well as other excerpts contributed by him.]

Dear Reader,

The idea for this pot pourri of letters, pseudo-articles, articles and poems was prompted by a convergence of chronological events going back some twenty years. In the first place I was confronted by imminent retirement after an uninterrupted thirty-year period of full-time research and secondly, an anniversary was approaching of a description of the liposome model as we now know and use it, for which I was partly responsible. The formulation was devised from a feeling that too much of our published scientific writing was stereotype, banal, platitudinous and devoid of all sensibility to the fabric of life in scientific research, success or failure.

218

I thereupon wrote to some 50 scientists for whom, to my knowledge, the liposome model had become a part of life inviting them to write me an informal letter, poem or clerihew, providing it was not libellous and which would reach me by the 10th November 1982, the day of my retirement. I suggested that they should chronicle an anecdote or observation, a hypothesis or comment about liposomes and/or the manner in which these delightful little objects had interfered with their lives.

For those of you who know nothing about liposomes or even from which era of biological research this volume springs, let me sketch the state of our knowledge of biological membranes some twenty years ago. By 1960 most biologists were attuned to the concept proselytized by Davson and Danielli that cells were bounded by a continuous bimolecular film or 'membrane' of phospholipid molecules arranged back to back. Robertson had 'visualized' such an anatomical structure not only surrounding the periphery of the cell (the plasma membrane) but also around many intracellular organelles. In December 1961 Rudin, Mueller, TiTien and Westcott announced that they had actually physically separated two aqueous compartments by a macroscopic (1—10 mm$^2$) membrane of a thickness of not more than two phospholipid molecules; the physical properties of their model were truly amazing, particularly to biologists. Starting in November 1961, shortly after the arrival of the first EM at Babraham, Horne and I attempted to visualize dispersions of phospholipids in aqueous negative stain. Towards the end of 1962 we had persuaded ourselves that we were seeing minute sacs of approximately 50 nm diameter, the first 'lipid somes' as we have come to know them. Not until 1964 had we established beyond reasonable doubt the integrity of these spontaneously formed, closed membrane systems, tantalisingly similar to the membranes which appeared to surround cells. These informal letters recapitulate the development and diversification of the liposome 'membrane' model from the early days up to the present.

I would agree with James Watson that '...science seldom proceeds in the straightforward manner imagined by outsiders. Instead, its steps forward (and sometimes backward) are often very human events in which personalities and cultural traditions play major roles' (Preface, *The Double Helix*, Weidenfeld and Nicolson 1968). The letters which follow testify to Watson's sentiment and to my mind, reflect an overall insight into the business and pleasures of scientific research.

For the benefit of prospective editors of future scientific volumes and for the enlightenment of those who have suffered the frustration of delayed publications in the past, I present below a histogram of the actual months of arrival dates of promised communications. I feel justified in releasing such data not because I think any of the present contributors are at this moment exuding adrenaline but rather because of the statistical significance of such a uniquely large number of contributors for what is, after all, a small book. The volume was not held up, I simply ignored dilatory correspondents and acknowledged those who owned up that they had no stomach for a script! Unquestionably most of the correspondents enjoyed the exercise.

**A D Bangham**

```
                                     x
                                     x
                                     x
                                     x
                                     x
                                     x
                                     x    x
                                     x    x
                                     x    x    x    x
                                     x    x    x    x
                                     x    x    x    x
                                x    x    x    x    x
                                x    x    x    x    x    x
            x                   x    x    x    x    x    x
Mar  Apr  May  Jun  Jul  Aug  Sep  Oct  Nov  Dec  Jan  Feb  Mar
          1982                                   1983
```

## Virus infection of liposomes (or virology for the microveterinarian)

*Anne M Haywood*

Departments of Pediatrics and Microbiology, University of Rochester, Rochester, NY 14642 U.S.A.

It has been a decade since I first wandered into your laboratory with the intent of finding out whether liposomes were susceptible to viral infections, and if so, could I find out enough about the molecular events during infection to find a way to cure them. It seems a good time to review what has happened with this microveterinarian (or is it minipediatric) undertaking.

Perhaps we have known each other long enough now to tell you how I first became interested in liposomes. Back in the late 1960's when I was working on RNA bacteriophage, RNA and protein synthesis, I found myself one sleepy rainy Saturday afternoon curled up in a big comfortable chair in the library nodding over the RNA literature until I finally dropped the journal on the floor. When I picked it up, it fell open to a paper by Bangham, Dingle and Dame Honor Fell at the Strangeways Laboratories. Since puns were a way of life, I almost could not believe my eyes, but when I stopped laughing, I started to read and then I moved on to your liposome review. So that is how I got hooked on your work and liposomes.

\*\*\*\*\*\*

## Reconstitution of membranes

*Efraim Racker*

Section of Biochemistry, Molecular and Cell Biology, Division of Biological Sciences, Cornell University, Ithaca, New York 14853

[Dr Racker interspersed a historical account of his work with liposomes with these aphorisms.]

**Lesson 1.**
If you work on a system of multifactorial factors be prepared to make multifactorial mistakes. You all know about Vitamins $B_1$, $B_2$, $B_6$, and $B_{12}$. What happened to $B_5$, $B_8$, $B_{11}$, etc.? The road to a successful analysis of a complex multifactorial system is like Piet Hein's 'Road to Wisdom.'

> *'It is plain*
> *and simple to express*
> *Err, err and err again*
> *But less and less and less'*

**Lesson 2.**
Don't force your postdoctoral fellows to listen to your brilliant ideas. Make your suggestions as persuasive as possible. I say to them: 'These experiments are so exciting that if you don't want to do them I will do them myself.' This policy had two advantages:

(*a*) some postdoctoral fellows did get excited and tried the experiments;

(*b*) others did not and have thus kept me busy in the laboratory for the past 30 years.

**Lesson 3 (Chris Miller). Early decisions**
If you don't find the activity you want, study the activity you get; it's probably more interesting than what you had expected.

**Lesson 5. After you publish**
Rejoice when other scientists don't believe what you know to be true. It will give you extra time to work on the phenomenon in peace. When they start claiming that they have discovered it before you, look for a new project.

222

# How liposomes influenced my life and got away with it

*Gregory Gregoriadis*

Clinical Research Centre, Harrow, Middlesex, HA1 3UJ, UK.

*Albert Einstein, The Bronx, New York.*
*Gentiles and Jews (impartial to pork),*
*academic discussions and ifs and buts*
*ceruloplasmin and test tubes, decapitated rats.*
*Exciting sequel of a previous discovery.*
*One could deduce from the total recovery*
*of counts injected by the parenteral route*
*into the blood stream or the pad of the foot:*
*desialylated proteins home to the liver!*
*Will drugs be targeted? Post doctoral fever...*
*November of sixty-nine, still in New York,*
*expiring visa, in search of work.*
*Abandoned reagents of forgotten adventure,*
*unfinished note books. An issue of Nature,*
*classified advertisement, turn of fate*
*enzyme entrapment, tantalizing bait!*
*Dear Madam of London U.K.*
*I am flying to join you, if I may.*

*Little fatty vesicles of bilayer fame*
*protean and elusive, fragile all the same,*
*aloof and enigmatic beneath your many skins,*
*unyielding to the rigour of thousands of spins,*
*descended from the pastures of Babraham I am told,*
*you never ceased to wrinkle, expand and then to fold*
*embracing sodium ions and such electrolytes.*
*Twinkling guide stars to throngs of acolytes*
*desirous of your membranous semi-barriers.*
*Precursors of bion, potential drug carriers.*

*Tasty and soft, milky and fat*
*trapped in the veins of gnashing rat,*
*trembling and pale carrying their load*

223

*of lytic enzymes along the road,*
*running and tumbling squeezing through membranes*
*frantic blind donkeys with disheveled manes,*
*crumbling frail travesties of wily Trojan horses*
*swallowed by scavengers through endocytosis.*
*Some went astray in the midst of confusion.*
*Tales of adsorption, or was it fusion?*
*Lamellar vesicles unstable, short-lived*
*whether uncharged or meticulously sieved,*
*doomed to decay in looming Kupffer cells,*
*acidic enclaves, milieu that quells*
*hydrogen bonds, biopolymer tresses.*
*Ossified preys of Harpean caresses.*


*Further experiments out in Harrow*
*on vesicle fate in spleen and marrow,*
*the carriage of antigens for potent vaccines,*
*of killer molecules, on novel means*
*to alleviate models of storage disease.*
*Meandering ruses to quench or appease*
*the lust of reticular prurient ogre*
*for fatty vesicles a little longer,*
*the piercing fury of blood components,*
*hyper-dense globins and other opponents.*
*Snatching Scyllas forced to recoil,*
*vascular cosmos in muted turmoil.*


*Victorious vesicles sturdy and lasting*
*briefed to target by way of casting*
*cytophilic ligands on outer boundaries,*
*hydrophobically soldered. Cerebral foundries*
*of dwarfish harriers made to tease*
*bewitching Sirens and 'scape with ease.*
*Compulsive dreams of toxophilite*
*fostered, I remember, by sagacious Knight,*
*inspired together with scores of zealots,*
*fulfilled by willing unbridled Helots.*

*Anxious years of soul searching,*
*of doubts, frustration. Memory etching*
*with ghosts of mice, rabbits and apes,*
*with public debates, sour grapes,*
*enchanting persons from five continents,*
*sleek orators, verbal incontinence,*
*nightmarish repeats of 'next slide',*
*recent data to disclose or hide.*
*Enthralling years gone and to come,*
*emaciated time perhaps to some?*
*O Delphian sibyl answer me. I need*
*to know. Pythia, tell me. 'Succeed*
*thou shalt not for thee defeat in battle'*
*Thus spoke the oracle, cruelly subtle...*

# English is our second language

*Nicholas P Cristy*

RBP from *New England Journal of Medicine* **300** No 17, 26 April 1979, p 979 (condensed).

Imagine a medical student, graduated two years ago with a degree in science or the humanities, perhaps with high honors. In his third or core clinical or major medical year, not yet having forgotten all that he learned in college, eager to apply his new 'preclinical' erudition to patients, he has just started his clerkship in internal medicine. He goes on rounds. He goes on rounds again. And again. Something is wrong. He understands nothing of what is being said. He understands well enough the scientific material that flooded him in the first two years of medical school, but he can grasp very little now. After about a week he sees the trouble. What is spoken on rounds is not English.

This discovery brings with it immediate relief. All that is necessary is to find out what language is spoken, what its name is, what is its grammar, what are its idioms, above all, what are its purposes. Like any other traveler in a foreign country, the freshman clinical clerk begins to absorb this new language and soon has slid with the ease of youth into its usages, repellent though they may have seemed at first. He divines its aims. He learns it so well that when the clerkship is over, his facility may surpass his teachers'. His linguistic thrusts, after three months of daily practice, often leave his instructors gasping—and uninformed. The student has learned the passwords. He knows Medspeak.

## Medspeak

Medspeak is an Orwellian invention of interns and residents, a lingua franca: 'a spoken language used for communication among speakers of different and mutually unintelligible languages ... used solely for limited purposes.' [1] By this definition, Medspeak can be learned by anyone, even a medical student, just as anyone can learn Malay, Swahili, the commercial language of the Mediterranean littoral or the pidgin (business) English of the Chinese coast. The key words are 'communication' and 'limited.'

'Communicate' means 'give to another as a partaker, impart, transmit', to exchange thoughts; the Latin original means 'shared', a popular word in the social sciences today. How much giving and exchanging, how much transmitting and sharing can be done in Medspeak? Very little. The communication is limited. Here are some of the reasons.

*The need to appear learned*
Michael Crichton has analysed this common problem in medical writing [2]. The disease is well established in Medspeak too. The signs are many. Most obvious is the use of a big word when a small word will do. 'Symptomatology' is particularly good. It has six syllables. It has Greek origins. It is suitably cacophonous. It implies familiarity with Medspeak. Best of all, nobody knows what it means, not even dictionaries.

Another is 'armamentarium', also hexasyllabic but more humbly derived from Latin. But how appropriately

226

derived—from the language of the warlike Romans. One imagines Caesar's legions with their baggage trains and arsenals and catapults. What a good word to convey the idea of an attack on patients. High-sounding, formal, frightening. A fine, big word.

Medspeak also encourages the use of several words when one will serve [2]. 'Prior to' is better than 'before' The same student, the same resident who would never utter, 'Prior to the time I met you I didn't know what love was', does not blush to say on rounds, 'Prior to admission ... (PTA)' in every third sentence; or 'Prior to the hemoptysis there was dyspnea.' In English, 'before is simpler, better known, and more natural, and therefore preferable'. [3] More natural in English, but we are speaking Medspeak, we are being learned.

### The need to be brief

Since so much time on rounds is taken up with those big words and extra words, one has to cut corners somewhere. The right corrective is to abbreviate. Ambiguities may result, but the thing is to save time. There are several more patients to see. 'This woman is to have PT OD.' Will a daily venipuncture be done for measurement of prothombin time or will she be taken five times a week to physical therapy? It makes a difference, but we must hurry. Let us hope the order book is clearer. 'We had a DOA last night.' That's not so bad. A Dream of Avarice is not a mortal sin. 'He entered in a bad DKA.' Did the patient drive into the emergency room in a damaged German sports car? He did not. He had diabetic ketoacidosis. Time is short. Yes, so 'How is this patient progressing?' 'Zero delta', the intern replies. Perfectly good Medspeak. Not the usual form or acrostic abbreviation but an acceptable idiom, known as Medspeakeasy. 'Zero delta' means of course 'no change' in English, but again, we are not speaking English. Let's move on. 'Oids' is another specimen of Medspeakeasy. 'What are you giving this lady?' 'Oids.' 'Oids' means '-oids' but even in Medspeakeasy it is hard to say the hyphen. You just have to know. 'Oids' stands for steroids—corticosteroids.

The above are not inventions. They are real. They happen every day and do their work. The attending is ashamed to ask what they all mean, so they save time.

*The need to hide*

From his associates on the service team, the new clinical clerk soaks up this craving with dreadful speed. Within days all the new young people making rounds have become invisible, hidden behind their verbal screens. Medspeak is rich in screens. It may be poor in simple nouns and verbs and in short, active, declarative sentences, but what of that? With screens like these who needs words? In Medspeak 'Occasionally' sometimes means 'three times a week', sometimes 'varying between three times a week and three times a year.' Sometimes it signifies 'not very often', sometimes 'very rarely', sometimes 'sometimes'; sometimes 'I don't know.'

'Not really' is a fine screen, too; not a coarse one, a fine one. Its meaning is impenetrable. 'Was this man anemic PTA?' (there is our elegant friend 'prior to' again, hiding behind an abbreviation). Answer: 'Not really.' Does this mean the hematocrit ('crit' in Medspeak) was 35 or 65 per cent? Anything is possible.

When the residents use Medspeak's cardinal screen, it has everything. It is Latinate. It has four syllables. It has as many meanings as the hydra has heads. Its name is 'essentially.' In his real life, would the student or intern ever answer the question 'Do you like French cooking?' with 'Essentially, no'? Indeed he would not. Only in the medical environment does his good sense desert him and his Medspeak take control. There, the meanings of 'essentially' stretch before him without limit. Those meanings? 'Precisely', 'imprecisely', 'wholly', 'partially', 'quite', 'almost', 'not.' The resident reads from a real-life laboratory slip reporting toxicologic findings in a case of overdose with multiple but unknown drugs: 'Barbiturate level essentially normal.' Normal for a normal person? Normal for a person who has taken an overdose of barbiturate? Normal for a person who has taken an over-dose of something other than a barbiturate? Or a little above or a little below 'normal' for any or all of the above? A triumph of Medspeak. Meaning is drained out, communication ceases to exist, thought wilts, action is paralysed. The screen has screened. The young doctors are hidden and so is the laboratory. The attending is left alone to stare, helpless and immobile, at the nullity around him.

## Medspeak versus English

The consequences of Medspeak—that is, the consequences of pedantry, criptic brevity and the use of verbal smoke screens—are funny, so long as communication is not the purpose of spoken medical language. We see that the purposes of Medspeak are not communicative but manipulative, like Mediaspeak, and perhaps derived from Mediaspeak. Those purposes are to look smart, to get on with the job, to control the attending physicians and to avoid making a definite commitment to a diagnosis or a course of action. Aristotle, who wrote more than 2300 years ago, probably qualifies as an old fogy, 'an old-fashioned or excessively conservative person'; he says in the *Rhetoric* that language is for communication: 'We may therefore regard it as settled that a good style is, first of all, clear. The proof is that language which does not convey a clear meaning fails to perform the very function of language.' [1] An outmoded point of view, but, if we were to espouse it again, we would get on with the work more expeditiously; everyone would know, more or less, what is going on, and we might all learn something from rounds.

As teachers of medicine, without being schoolmarmish or cranky, we owe it to the novice clinical clerk and the house staff to keep language from being corrupted, and thought thus blunted. Clinical medicine is ambiguous enough without cloaking it further in the cited imprecisions of speech. Left to ooze on toward unintelligibility,

Medspeak will end up being spoken in grunts and semaphores and will have achieved its final aim: to be 'a hybrid, partially developed language used solely for limited purposes.' [1] Now is as good a time as any to recover English as our first language, the one we were born with, and relegate Medspeak to second place. In time and with luck it might disappear, so that, at the very least, we can talk to one another [2, 5].

*References*

[1] *The Columbia Encyclopedia*. Second edition. New York, Columbia University Press, 1950, p 1139

[2] Crichton M *Medical obfuscation: structure and function.* N. Engl. J. Med. 293:1257–1259, 1975

[3] Gowers E *The Complete Plain Words*. Baltimore, Penguin Books, 1962, pp 192–193

[4] Cooper L *The Rhetoric of Aristotle*. New York, Appleton-Century-Crofts, 1932, p 185

[5] Christy NP *The twenty-seventh anniversary and some thoughts on isolation.* J. Clin. Endocrinol. Metab. 27: 1778–1783, 1967

# Next slide please

*David Davies*

RBP from *Nature* **272** 743, 27 April 1978. © 1978 Macmillan Magazines Ltd.

I thought that in the eight minutes I've got I'd bring you up to date on what our group has been doing in the last year; in a sense this is a progress report and updates the paper we gave here last year; I won't go over the nomenclature again; could I have the first slide please—oh, I think you must have someone else's box—mine is the grey one with my name on the top, no, wait a minute, not my name, whose name was it now? ah yes, you've found it; there's a red spot on the top right hand side of each slide that is the side that becomes the bottom left when you project it. OK, you've got it now, let's have a look, no, that's the last slide not the first, yes, now you're got the right one but it's on its side, what about the red dot? there are two? well anyway turn it through ninety degrees, no, the other way, yes now we're there, perhaps we could have the lights off, well I'm sorry there are probably too many words on this slide, and the printing is a bit thin; can you read it at the back? you can't; well I'd better read it out; no I won't, it's all in the paper which should be published within a month or so, and anyone who wants I'll give a preprint to afterwards, anyway, for those who can read it, this slide is a block diagram of the purification process we used and before I go any further I should mention that there are a couple of misprints: on the third row, fourth box from the left, well, of course that's the second box from the right, if you can read it, it says alkaline, now that should be acidic; also you can perhaps see the word mebmrane, that should of course be membrane; now if I can have a look at the next slide—now which one is this? ah, yes it's the scatter diagram. I haven't marked the quantities but we are plotting concentration against particle size; if I remember rightly this has been normalised; perhaps I could have the lights for a moment to check in the text, yes, here we are, well it doesn't actually say—we could work it out but it's probably not worth the time, so if I could have the lights off, let's have a look at the plot; well I think you can see a sort of linear relationship—there's a fair bit of scatter, of course, but I think the data are at least suggestive; perhaps if I held up a pointer you could see the relationship more clearly—I expect there's a pointer around somewhere, no I won't need the lights, yes here it is, now you can see the trend and there's just the hint of another trend running subparallel to it through this other cluster of points, you may see that more clearly if I slide the pointer across to the other—no, I wasn't saying next slide, just that I would slide the pointer; anyway now the next slide is up let's keep it on the screen, now this is the sort of evidence on which the data in the last slide were based; this is a thin section—it could take just a bit of focusing—yes, that's better, it's difficult to get the whole slide in focus at once, now the scale is, well that bar is one micron long, hang on what am I saying? it's ten microns long—oh dear, the chairman is giving me the two minute warning, it's dif-

Next slide please

ficult to give you a clear picture of this work in only eight minutes, but let's plough on, what was I saying? ah yes, that bar is ten microns long, now if we turn to the next slide, please, this is the result of a chemical analysis of the dark region that is near the centre of that thin section, is it possible to go back a slide? well not to worry, you can see in the analysis how dominant—sorry what was that? oh yes, the errors are plus or minus a per cent or so—that's the standard deviation, no it can't be, it must be the standard error of the mean—oh dear, the chairman says my time is up, can I beg half a minute—are there any more slides? really? well let's skip the next two, now this one is pretty important, it brings together several of the threads that you've probably been able to discern running through this talk, but rather than go through it in detail perhaps I should have the lights and just put up one or two key numbers on the blackboard—the chairman says there's no chalk, well it's all in the paper I was mentioning anyway perhaps I've been able to give you the gist of what we've been doing, I guess that's all I've got time for.

## Two famous papers

*Peter Elias*

Reproduced from *Transactions on Information Theory* vol IT-4, 3 September 1958, p 99. © Institute of Electrical and Electronic Engineers, Inc.

It is common in editorials to discuss matters of general policy and not specific research. But the two papers I would like to describe have been written so often, by so many different authors under so many different titles, that they have earned editorial consideration.

The first paper has the generic title 'Information Theory, Photosynthesis and Religion' (title courtesy of D A Huffman), and is written by an engineer or physicist. It discusses the surprisingly close relationship between the vocabularly and conceptual framework of information theory and that of psychology (or genetics, or linguistics, or psychiatry, or business organization). It is pointed out that the concepts of structure, pattern, entropy, noise, transmitter, receiver, and code are (when properly inter-preted) central to both. Having placed the discipline of psychology for the first time on a sound scientific base, the author modestly leaves the filling in of the outline to the psychologists. He has, of course, read up on the field in preparation for writing the paper, and has a firm grasp of the essentials, but he has been anxious not to clutter his mind with such details as the state of knowledge in the field, what the central problems are, how they are being attacked, et cetera, et cetera, et cetera.

There is a constructive alternative for the author of this paper. If he is willing to give up larceny for a life of honest toil, he can find a competent psychologist and spend several years at intensive mutual education, leading to productive joint research. But this has some disadvan-tages from his point of view. First, psychology would not be placed on a sound scientific base for several extra years. Second, he might find himself, as so many have, diverted from the broader questions, wasting his time on problems whose only merit is that they are vitally important, unsolved, and in need of interdisciplinary effort. In fact, he might spend so much time solving such problems that psychology never *would* be placed on a sound scientific base.

The second paper is typically called 'The Optimum Linear Mean Square Filter for Separating Sinusoidally Modulated Triangular Signals from Randomly Sampled Stationary Gaussian Noise, with Applications to a Pro-blem in Radar.' The details vary from version to version, but the initial physical problem has as its major interest its obvious nonlinearity. An effective discussion of this problem would require some really new thinking of a dif-ficult sort, so the author quickly substitutes an unrelated linear problem which is more amenable to analysis. He treats this irrelevant linear problem in a very general way, and by a triumph of analytical technique is able to present its solution, not quite in closed form, but as the solution to an integral equation whose kernal is the solution to another, bivariate integral equation. He notes that the problem is now in a form in which standard numerical analysis techniques, and one of the micromicrosecond computers which people are now beginning to discuss, can provide detailed answers to specific questions. Many authors might rest here (in fact many do), but ours wants real insight into the character of the results. By carefully taking limits and investigating asympotic behavior he succeeds in showing that in a few very special cases (which include all those which have any conceivable application or offer any significant insight) the results of this analysis agree with the results of the Wiener – Lee – Zadeh – Raggazzini theory—the very results, indeed, which Wiener, Lee, Zadeh, and Raggazzini obtained years before.

These two papers have been written—and even published—often enough by now.

I suggest that we stop writing them, and release a large supply of manpower to work on the exciting and important problems which need investigation.

# The conference

*Douglas E Kidder*

RBP from *Verses Bright and Beautiful* University of Bristol, 1983

Academics are sometimes believed to spend much of their time going to conferences in distant and glamorous places at somebody else's expense. Perhaps a few manage to do this, and so perpetuate the legend. Of course, as most academic establishments, at least in this country, can only afford to have one specialist in any one field, some travelling is necessary if we are to discuss our work with other people who have the same problems. Unfortunately, the expenses available rarely quite stretch to the cost, and still more unfortunately, our meetings are more likely to be in Birmingham than Hawaii. However, in the course of our work, we may with luck get one or two memorable visits of a lifetime, which is more than an assembly line worker can hope for.

234

**Twinkle, twinkle, learned Prof**

*Twinkle, twinkle, learned Prof,*
*How I wonder where you're off,*
*Up above the world, so high,*
*in a Jumbo in the sky —*

*To some conference afar,*
*Where they see you as a star,*
*Showing, with your mastermind,*
*Work by suckers left behind.*

*May you be a shining light!*
*Life's worth while when things go right.*
*May the questions never go*
*Way beyond the stuff you know!*

*So, good fortune when you fly*
*in your Jumbo, up so high,*
*Though we don't know where you're off,*
*Twinkle, twinkle, learned Prof!*

---

## Technical expressions

David Kritchevsky, *Medical Tribune* 3 August 1964.

### (a) Conference Discussions

| When They Say: | They Mean |
|---|---|
| The two sets of data are roughly comparable. | What's 100 per cent difference between friends? |
| I present these data for your consideration and guidance. | Any ideas? |
| I apologize for this slide. | The data are too poor to present clearly. |
| I'll discuss this later with anyone who is interested. | I can worm out of it more neatly in private. |

| | |
|---|---|
| We haven't done it yet, but it's on the books. | FIRSTIES! |
| We are not yet in a position to publish this. | No consistent results. |
| I'm not sure I understand the question, but let me say this... | You keep going your way; I'll keep going my way. |
| I'm a little worried about your last statement. | I'm not, really—but you had better be. |
| Thanks for your comments. | Same to you, fellah! |
| We don't take a very strong position. | We did until your last question. |
| Animal studies were equivocal, but our tissue culture experiments show... | *In vitro veritas!* |
| May I make a brief comment? | (Six slides). |
| I don't think we'll take this approach... | But you all heard who said it first. |
| As you and I have both shown... | Let's be buddies again. |
| I'm indeed sorry that Dr X couldn't be here to discuss this point... | Like hell I am. |
| I'm sorry that Dr Y isn't here today to answer this question... | I *must* remember to bring the junior people with me. |
| We started with rather large quantities because of the intricate synthesis | We're pretty sloppy, so we use a lot of material to start with. |

|  |  |
|---|---|
| Control studies show... | These were carried out in an adrenalectomized, hepatectomized, gonadectomized, hypophysectomized, otherwise normal animal. |

******

### (b) Federalese

G R Hicks. RBP from a 1956 issue of *Word Study*, © 1956 by Merriam-Webster Inc., publisher of the Merriam-Webster dictionaries.

Channels = The trail left by interoffice memos.

Consultant (or Expert) = Any ordinary guy more than 50 miles from home.

To activate = To make carbons and add more names to the memo.

To implement a program = To hire more people and expand the office.

To clarify = To fill in the background with so many details that the foreground goes underground.

We are making a survey = We need more time to think of an answer.

Note and initial = Let's spread the responsibility for this.

Will advise you in due course = If we figure it out, we'll let you know.

Referred to a higher authority = Pigeonholed in a more sumptuous office.

Research work = Hunting for the guy who moved the files.

Further substantiating data are necessary = We've lost your stuff. Send it in again.

We're exploring the problem = Don't get impatient. We'll think of something.

******

D J Harris, Ship Division, National Physical Laboratory. *NPL News* **205** 15, 21 May 1967.

### (c) Hydrodynamic Definitions

| | |
|---|---|
| Naked resistance | Reluctant nudist |
| Skin friction | I was in Lux |
| Total drag | Utter bore |

| | |
|---|---|
| Duct | Lowered the head hastily |
| Added lift | Penny a mile extra |
| Beam/length ratio | 36/68 |
| Propeller | To push one's female companion |
| Turbulent boundary layer | Angry groundsman |
| Hovercraft skirt | Typist at H.D.L. |
| Hump speed | Fast camel |
| Streamline flow | 26/26/26 |
| Total resistance | Spoil sport. |

# How to survive a conference

*Brian McEnaney*

From the 'Lateral Thoughts' section of *Physics World*, May 1989 (Bristol: Institute of Physics Publishing Ltd).

Most scientists are like most athletes—very few athletes have the physical prowess of an Olympic gold medallist and very few scientists have the intellectual gifts of a Nobel Laureate. Yet in science, unlike athletics, it is often necessary for scientists who are hardly up to club standard to perform in the same event as the international star. That event is the scientific conference and, to survive it, the mediocre or merely competent scientist must develop skills which would be redundant to a Fellow of the Royal Society.

Naive scientists often think that the object of conferences is to report their latest results and so make a contribution to the development of their subject. This is quite wrong. Conferences are not science—they are theatre. The object is not to exchange ideas and knowledge, but to dazzle your audience during your presentation and amuse them afterwards by outwitting your questioners. Brilliant scientists do this with lordly ease, but, to survive in such company, the mediocre scientist must resort to less noble stratagems.

238

The tactic most likely to ensure survival is the brazen delivery. Even if you have secret doubts about the validity of your results and your own competence, you must present your work with an air of impregnable confidence. A scientific audience is like a pack of wild animals—they can smell fear at a hundred yards—and once they sense nervousness or uncertainty, they show no mercy. Recognising this, you must become a lion tamer. Adopt a loud, confident delivery, perhaps even with a hint of menace. Your audience will be suitably cowed and reluctant to press you with searching questions. Also, as a brazen author, you should never apologise or make excuses for the few results and the Mickey Mouse analysis that you are presenting.

Your audience are not interested in your problems and excuses—they have problems of their own. Perhaps the Board of Directors are considering closure of the Research Department, or the Minister of Education has refused them a pay rise, or they are going through an expensive divorce. Compared to all this, your six-month wait for samples that turned out to be the wrong material is very small beer.

An alternative to the brazen strategy is the disarming approach. A favourite device of disarmers is to present their work as preliminary results. In this way they hope that they will not be subject to such searching scrutiny as would be the case if they were presenting a finished piece of work. There are many experienced scientists who employ this tactic, even though they have been working on the same topic for up to 20 years. If you remain incurably nervous and dubious about your results, remember, in this age of scientific specialisation, the chances are that there are only two other people in the audience who are competent to judge your work; these are your co-authors and it is clearly not in their interests to give the game away.

Many scientists are less nervous of presenting their work than of answering questions in the following discussion. Provided that a few of the disarmers' stratagems are mastered, this need not be a traumatic business. A favourite tactic is to ask the chairman to re-arrange the session so that your paper is last. Most session chairmen are unable to keep up the schedule, so with luck and over-runs, there will be no time to discuss your paper before the coffee break. However, if you must answer questions, do not become aggressive. If some old duffer points out that your approach merely duplicates his classic work with Waffle and Wordmincer in 1952, except that they came to exactly the opposite conclusion, it is completely the wrong response to leap off the podium and wrestle him to the floor or in some other fashion to lose your cool. A useful rejoinder is to ask your inquisitor to repeat the question as you are slightly deaf. This has two benefits: it gains the sympathy of the rest of the audience and gives you time to think of a reply. A good all-purpose response to comments of this type is: 'That is the traditional view of this subject, but more recent work points in a completely different direction.' Notice that you have not actually contradicted him, but you have managed to imply that he has not read the literature for 30 years. Would-be disarmers should note that the sympathy card must be played with discretion. Wearing dark glasses and having the session chairman lead you to the podium with crutches is probably going slightly over the top.

A common problem at conferences is the persistent questioner. If the session chairman will not shut him up, then it is up to you. A good way is to say: 'Well that is a rather detailed and complex question, which we might better discuss over coffee.' Of course, when coffee time comes, you make yourself scarce. Another difficult type of comment is along these lines: 'I am surprised that you used polycrystalline samples. Watter, Rott and Song showed in 1956 that polycrystalline samples give irreproducible results and all work since then, except yours, has used annealed single crystals'. With an inexperienced author this type of comment often produces an awkward pause. But the practised disarmer will immediately reply: 'I am sorry there are matters of commercial confidence associated with this work which I am unable to talk about at present.' Not only does this stop the line of questioning dead, it worries the opposition in case you are on to something they have missed.

With a judicious combination of such tactics, the mediocre scientist can not only survive the conference, but also enhance his reputation. But, if your object is merely to survive, beware of overdoing it. If you give too good a performance, you will be invited to give a plenary lecture at next year's conference.

---

# Fluorescent dyes for differential counts by flow cytometry: does histochemistry tell us much more than cell geometry?

[Dr Howard Shapiro chose an interdisciplinary format for his serious study on fluorescent dyes: accompanying himself on the guitar he sang the paper at the 28th Annual Meeting of the Histochemical Society in Chicago.]

RBP of Howard M Shapiro *Journal of Histochemistry & Cytochemistry* vol 25 No 8 pp 976–89 (1977).

Blood cells are classified by cell and nuclear shape and size
And texture, and affinity for different types of dyes,
And almost all of these parameters can quickly be
Precisely measured by techniques of flow cytometry.

It's hard to fix a cell suspension rapidly and stain
With several fluorochromes, and this procedure, while it plain-
Ly furnishes the data which one needs to classify,
May fade away, and newer, simpler, methods never dye.

The stains which hematologists routinely use on smears
Assiduously kept their basic nature through the years
Since Romanowsky made his variation on a theme
Paul Ehrlich wrote some time before he chased the treponeme.

FLUORESCENT DYES FOR DIFFERENTIAL COUNTS BY FLOW CYTOMETRY: DOES HISTOCHEMISTRY TELL US MUCH MORE THAN CELL GEOMETRY?

The stains which he·ma· to·lo·gists rou· tine·ly use on smears as-
sid·u·ous·ly kept their ba·sic na·ture through the years, since
Ro·ma·now·sky made his va·ri· a·tion on a theme Paul
Ehr·lich wrote some time be·fore he chased the tre·po· neme.

(music © Howard M. Shapiro - used by permission)

**Figure 1.** Melody line and chord harmony for oral presentation of this work (Howard M Shapiro, M.D.; used by permission).

When acid dyes and basic dyes were mixed, one could obtain
What Ehrlich, half erroneously, called a neutral stain,
Which helped him classify the leukocytes, or other cells,
By different colorations in their different organelles.

A further differentiation of cell types was made
By using several acid dyes, each of a different shade,

But then, a Russian's giant step stopped further progress cold,
When Romanowsky's methylene blue jar was left to mold.

In Romanowsky's paper, Leishman's, Giemsa's, and in Wright's,
The accent was on staining of malarial parasites;
Invisible plasmodia could now be seen for sure,
In blood cells stained with eosin and methylene azure.

Since white cells could be classified by Romanowsky's stain,
The mixture and its variants became, and still remain,
The basis of the differential counts technicians do,
And of some automated differential counters, too.

Computer image processing's an old familiar theme
To many engineers, who see the blood as heme sweet heme,
And nothing seemed more normal than to scan a smear and try
To duplicate the differential as it's done by eye.

From digitized, scanned images of blood cells on a slide,
Key features are extracted; cells are then identified
From distributions of a few—say four or more—of these,
By multivariate statistical analyses.

With this in mind, it shouldn't come as much of a surprise
That color, granularity, and cell and nuclear size
Are useful as parameters for programs which decide,
In much the way that Ehrlich did, how cells are classified.

The methods of photometry which Ehrlich could obtain
Were nowhere near a match for one good eyeball and a brain,
But cytophotometric apparatus, nowadays,
Can help us class the leukocytes in many different ways.

We've tried to automate the 'diff' to simplify the task
Of answering some questions hematologists may ask,
And now it looks as if they might soon get their answers out

Of instruments much simpler than the ones I've talked
about.

Some dyes are more specific than some other dyes, it's
true,
But more specific techniques may be harder ones to do.
A basophil count done by IgE receptor sites
Won't give you much more information than you'd get
with Wright's.

Now, surface antigens and lectin binding sites may tell
Us more about development and function of a cell;
With multiparametric methods, we will soon report,
One may work with cell mixtures and avoid the need to
sort.

Though flow cytometry has only come of age of late,
Its influence on modern cytochemistry is great;
If, in the future, we can look at it through Ehrlich's eyes,
We'll learn as much of drugs and cells as we have learned
of dyes.

*Acknowledgments*

My mother, Jennie E Shapiro, taught me, for a start,
Some histochemical techniques—the science and the art,
But I would not have shown my slides, nor strummed
upon my strings,
Without the help of other people, doing other things.

I thank Bob Young and Kathy Mead, V Hepp and Sarah
Lesher,
Who kept our apparatus fed with blood cells under
pressure;
Bob Webb and Henri Vetter, who made sure the beast
behaved,
And Joning Chan, who got the data analysed and saved.

# Mathmanship

*Nicholas Vanserg*

RBP from
*American Scientist*
**46** 94A (1958).

In an article published a few years ago, the writer [1] intimated with befitting subtlety that since most concepts of science are relatively simple (once you understand them) any ambitious scientist must, in self protection, prevent his colleagues from discovering that *his* ideas are simple too. So, if he can write his published contributions obscurely and uninterestingly enough no one will attempt to read them but all will instead genuflect in awe before such erudition.

## What is mathmanship?

Above and beyond the now-familiar recourse of writing in some language that looks like English but isn't, such as Geologese, Biologese, or, perhaps most successful of all, Educationalese [2], is the further refinement of writing everything possible in mathematical symbols. This has but one disadvantage, namely, that some designing skunk equally proficient in this low form of cunning may be able to follow the reasoning and discover its hidden simplicity. Fortunately, however, any such nefarious design can be thwarted by a modification of the well-known art of gamesmanship [3].

The object of this technique which may, by analogy, be termed *Mathmanship* is to place unsuspected obstacles in the way of the pursuer until he is obliged by a series of delays and frustrations to give up the chase and concede his mental inferiority to the author.

## The typographical trick

One of the more rudimentary practices of mathmanship is to slip in the wrong letter, say a $\gamma$ for a $\tau$. Even placing an exponent on the wrong side of the bracket will also do wonders. This subterfuge, while admittedly an infraction of the ground rules, rarely incurs a penalty as it can always be blamed on the printer. In fact the author need not stoop to it himself as any copyist will gladly enter into the spirit of the occasion and cooperate voluntarily. You need only be trusting and not read the proof.

### Strategy of the secret symbol

But if, by some mischance, the equations don't get badly garbled, the mathematics is apt to be all too easy to follow, *provided* the reader knows what the letters stand for. Here then is your firm line of defense: at all cost *prevent him from finding out!*

Thus you may state in fine print in a footnote on page 35 that $V^\alpha$ is the total volume of a phase and then on page 837 introduce $V^\alpha$ out of a clear sky. This, you see, is not actually cheating because after all, or rather before all, you *did* tell what the symbol meant. By surreptitiously introducing one by one all the letters of the English, Greek and German alphabets right side up and upside down, you can make the reader, when he wants to look up any topic, read the book backward in order to find out what they mean. Some of the most impressive books read about as well backward as forward, anyway.

But should reading backward become so normal as to be considered straightforward you can always double back on the hounds. For example, introduce $\mu$ on page 66 and avoid defining $\mu$ until page 86†. This will make the whole book required reading.

### The pi-throwing contest or humpty-dumpty dodge

Although your reader may eventually catch up with you, you can throw him off the scent temporarily by making him *think* he knows what the letters mean. For example every schoolboy knows what $\pi$ stands for so you can hold him at bay by heaving some entirely different kind of $\pi$ into the equation. The poor fellow will automatically multiply by 3.1416, then begin wondering how a $\pi$ got

---

†All these examples are from published literature. Readers desiring specific references may send a self-addressed stamped envelope. I collect uncanceled stamps.—N Vanserg.

into the act anyhow, and finally discover that all the while $\pi$ was osmotic pressure. If you are careful not to warn him, this one is good for a delay of about an hour and a half.

This principle, conveniently termed pi-throwing can, of course, be modified to apply to any other letter. Thus you can state perfectly truthfully on page 141 that F is free energy so if Gentle Reader has read another book that used F for *Helmholtz* free energy he will waste a lot of his own free energy trying to reconcile your equations before he thinks to look for the footnote tucked away at the bottom of page 50, dutifully explaining that what you are talking about all the time is *Gibbs* free energy which he always thought was G. Meanwhile you can compound his confusion by using G for something else, such as 'any extensive property.' F, however, is a particularly happy letter as it can be used not only for any unspecified brand of free energy but also for fluorine, force, friction, Faradays, or a function of something or other, thus increasing the degree of randomness, dS. (S, as everyone knows stands for entropy, or maybe sulfur). The context, of course, will make the meaning clear, especially if you can contrive to use several kinds of F's or S's in the same equation.

For all such switching of letters on the reader you can cite unimpeachable authority by paraphrasing the writing of an eminent mathematician [4]:

'When *I* use a letter it means just what I choose it to mean— neither more nor less ... the question is, which is to be master— that's all.'

### The unconsummated asterisk

Speaking of footnotes (I was, don't you remember?) a subtle ruse is the 'unconsummated asterisk' or 'ill-starred letter.' You can use P* to represent some pressure difference from P, thus tricking the innocent reader into looking at the bottom of the page for a footnote. There isn't any, of

---

* April fool. See what I mean?

course but by the time he has decided that P must be some registered trademark as in the magazine advertisements he has lost his place and has to start over again. Sometimes, just for variety, you can use instead of an asterisk a heavy round dot or bar over certain letters. In doing so, it is permissible to give the reader enough veiled hints to make him *think* he can figure out the system but do not at any one place explain the general idea of this mystic notation, which must remain a closely guarded secret known only to the initiated. Do not disclose it under pain of expulsion from the fraternity. Let the Baffled Barbarian beat his head against the wall of mystery. It may be bloodied but if it is unbowed you lose the round.

The other side of the asterisk gambit is to use a superscript as a key to a *real* footnote. The knowledge-seeker reads that S is $-36.7^{14}$ calories and thinks 'Gee what a whale of a lot of calories' until he reads to the bottom of the page, finds footnote 14 and says 'oh.'

## The 'hence' gambit

But after all, the most successful device in mathmanship is to leave out one or two pages of calculations and for them substitute the word 'hence' followed by a colon. This is guaranteed to hold the reader for a couple of days figuring out how you got hither from hence. Even more effective is to use 'obviously' instead of 'hence', since no reader is likely to show his ignorance by seeking help in elucidating anything that is obvious. This succeeds not only in frustrating him but also in bringing him down with an inferiority complex, one of the prime desiderata of the art.

These, of course, are only the most common and elementary rules. The writer has in progress a two-volume work on mathmanship complete with examples and exercises. It will contain so many secret symbols, cryptic codes and hence-gambits that no one (but no one) will be able to read it.

*References*

[1] Vanserg, Nicholas. How to write Geologese, *Economic Geology* vol 47, pp 220–3, 1952

[2] Carberry, Josiah. *Psychcoceramics* p 1167, Brown University Press, 1945

[3] Potter, Stephen. *Theory and Practice of Gamesmanship or the Art of Winning Games without Actually Cheating*, London: R Hart-Davis, 1947

[4] Carroll, Lewis. *Complete Works*, Modern Library edition, p 214

## How to set up a consortium

RBP from
S Verstov *Priroda*
No 7 (1981)
pp 127–8

Recently I received a reprint of an article, which I had co-authored, from a journal and decided to give it to my friend 'N'. We had studied in the same course and after graduation he had become a biologist and I a physicist. After several days I telephoned him and asked whether he liked the article.

'I didn't know that you wrote plays', said N. 'It's an interesting little novelty, only I don't understand why there are so many actors and how they relate to one another.'

'What play?' I bellowed into the telephone. 'That's a scientific article. You didn't read it through to the end.'

'N' stammered and admitted that he had not yet read the last page.

Then everything became clear. On the last page, as always, there was a description of the experiment and all the preceeding was occupied by a listing of authors. This is called a 'consortium', or, abroad, a 'collaboration', I explained to him.

My friend disapprovingly mumbled and hung up the receiver.

Soon, however, he himself called back and said that his colleagues were very interested in our method and asked me to tell them how they could also organize a consortium. I replied that I myself did not know how this was done but I could introduce him to the deputy scientific secretary of our institute who was in charge of these problems.

On the appointed day we appeared at the deputy's office.

'Will you tell me, please, how you organize a consortium?' asked 'N'.

'Oh, this is a very complicated and responsible job, I am not afraid to say—an art. First of all, it is necessary to construct an apparatus—as large a one as you wish. The bigger the apparatus the bigger the consortium will be, and the bigger the consortium, the more money they will give you for further development of the apparatus, since, in principle, this process is limited only by the total resources of the planet. That's why, by the way, international consortiums are popular now.'

'Obviously all those who constructed the apparatus also are members of the consortium?'

'Exactly the opposite. We have long understood that the more people who construct the apparatus the slower the work progresses. Usually we assign to this five or six people. When the apparatus is built and subsequently neglected it is necessary to organize the consortium. It is very important to choose the expedient moment. If the apparatus has not yet begun to operate then it is difficult to attract people, since there is no article which needs to be signed. If the apparatus has been in operation for a fairly long time it won't be considered new and then potential co-authors can be lost to a newer apparatus.'

'How do people find out that the apparatus is ready? Do you, indeed, attract as co-authors colleagues outside your own institute?'

'Of course, our goal consists in attracting a cadre from other institutes and laboratories. We have developed several methods of announcing the collection of authors. At first we simply posted announcements on the poles near the bus stops, however, this didn't prove entirely satisfactory since they telephoned us endlessly offering to swap two rooms. Then we began to print announcements in the evening paper. The result again proved insignificant, since it was discovered that the majority of scientific workers do not read newspapers. The most successful method, in my view, was the idea of sending our representatives to comradely dinners held upon the occasion of the defense of a dissertation or the awarding of a prize. After dinner, the guests are in a weakened state and readily agree to enter into the consortium.'

'Nevertheless, do you apply any sort of criteria in the selection of co-authors?'

'From my words you could have gotten the impression that anyone who wants to could become a participant in a consortium. Nothing could be more mistaken than this point of view. We adhere to a firm and categoric principle: every member of a consortium must, at least once, look at the apparatus. Such a rule permits him, in the future, not to confuse his own consortium with another. Earlier when we did not require this, some authors for years on end signed articles based on another apparatus. All of the consortiums were confused and it was necessary to create a commission which sorted out the authors according to their apparatuses.'

'An excellent principle. Permit me to write it down.'

'You can, of course, write down that the system is not perfect. For example, proceeding from this criterion, it was necessary to record as authors all members of delegations who visit our institute. Some delegates came from remote districts and regions and we had to for a long time search them out through the address bureau. Therefore we have now introduced a still stricter rule: every author must make at least one visit to the computation center. They don't take delegates there and hence the work with the authors is essentially regulated.'

'I am delighted with the clarity and logic of your criteria', declared 'N'.

'Now it is entirely clear to me how this is done. But how long can a consortium last and can the make-up of the authors be changed from time to time?'

'You have broached an extremely interesting question',

and, flattered by our high opinion of the activity, the deputy scientific secretary smiled. 'Recently we discovered an important law: any consortium is a self-perpetuating system. I have in mind the fact that the authors who have gone to see the apparatus or the computation center take their children with them since children in our times, as you yourself have undoubtedly noted, are extremely curious. They thus also become participants in the consortium. Now if an author is pensioned off, his children successfully continue to sign articles. However, don't think that we permit this process just to drift on. Recently we have introduced a third important restriction: every author must absolutely be literate. Thereby we resolutely exclude cases where some authors still do not know now to sign their names.'

When we came out of the office, my friend embraced me and shed a tear or two.

'You can't imagine how glad I am that I understand your system. Today I will describe it to our scientific secretary.'

In a few days he sent me a reprint of his latest article. I sincerely congratulated 'N' on his brilliant results. The listing of authors, was, of course, shorter than on our own articles but, nevertheless occupied three pages.

---

# How to display data badly

*Howard Wainer†*

[Methods for displaying data badly have been developing for many years, and a wide variety of interesting and inventive schemes have emerged. Presented here is a synthesis yielding the 12 most powerful techniques that seem to underlie many of the realizations found in practice. These 12 (the dirty dozen) are identified and illustrated. KEY WORDS: Graphics; Data display; Data density; Data-ink ratio.]

RBP from Howard Wainer. From the 'Commentaries' section of the *American Statistician*, May 1984, vol 38, No 2, pp 137–47

# 1. Introduction

The display of data is a topic of substantial contemporary interest and one that has occupied the thoughts of many scholars for almost 200 years. During this time there have been a number of attempts to codify standards of good practice (e.g., ASME Standards 1915; Cox 1978; Ehrenberg 1977) as well as a number of books that have illustrated them (i.e., Bertin 1973, 1977, 1981; Schmid 1954; Schmid and Schmid 1979; Tufte 1983). The last decade or so has seen a tremendous increase in the development of new display techniques and tools that have been reviewed recently (Macdonald-Ross 1977; Fienberg 1979; Cox 1978; Wainer and Thissen 1981). We wish to concentrate on methods of data display that leave the viewers as uninformed as they were before seeing the display or, worse, those that induce confusion. Although such techniques are broadly practiced, to my knowledge they have not as yet been gathered into a single source or carefully categorized. This article is the beginning of such a compendium.

The aim of good data graphics is to display data accurately and clearly. Let us use this definition as a starting point for categorizing methods of bad data display. The definition has three parts. These are (a) showing data, (b) showing data accurately, and (c) showing data clearly. Thus, if we wish to display data badly, we have three avenues to follow. Let us examine them in sequence, parse them into some of their component parts, and see if we can identify means for measuring the success of each strategy.

†Howard Wainer is Senior Research Scientist, Educational Testing Service, Princeton, NJ 08541. This is the text of an invited address to the American Statistical Association. It was supported in part by the Program Statistics Research Project of the Educational Testing Service. The author would like to express his gratitude to the numerous friends and colleagues who read or heard this article and offered valuable suggestions for its improvement. Especially helpful were David Andrews, Paul Holland, Bruce Kaplan, James O Ramsay, Edward Tufte, the participants in the Stanford Workshop on Advanced Graphical Presentation, two anonymous referees, the long suffering associate editor, and Gary Koch.

## 2. Showing data

Obviously, if the aim of a good display is to convey information, the less information carried in the display, the worse it is. Tufte (1983) has devised a scheme for measuring the amount of information in displays, called the data density index (ddi), which is 'the number of numbers plotted per square inch.' This easily calculated index is often surprisingly informative. In popular and technical media we have found a range from .1 to 362. This provides us with the first rule of bad data display.

*Rule 1—Show as few data as possible (minimize the data density)*
What does a data graphic with a ddi of .3 look like? Shown

**Figure 1.** An example of a low density graph (from SI3 (ddi = .3)).

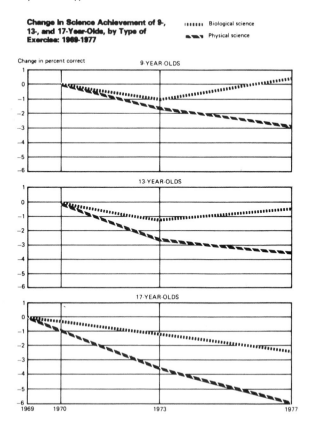

254

in Figure 1 is a graphic from the book *Social Indicators III* (SI3), originally done in four colors (original size 7" by 9") that contains 18 numbers (18/63 = .3). The median data graph in SI3 has a data density of .6 numbers/in$^2$; this one is not an unusual choice. Shown in Figure 2 is a plot from the article by Friedman and Rafsky (1981) with a ddi of .5 (it shows 4 numbers in 8 in$^2$). This is unusual for JASA, where the median data graph has a ddi of 27. In defense of the producers of this plot, the point of the graph is to show that a method of analysis suggested by a critic of their paper was not fruitful. I suspect that prose would have worked pretty well also.

Although arguments can be made that high data density does not imply that a graphic will be good, nor one with low density bad, it does reflect on the efficiency of the transmission of information. Obviously, if we hold clarity and accuracy constant, more information is better than less. One of the great assets of graphical techniques is that they can convey large amounts of information in a small space.

We note that when a graph contains little or no information the plot can look quite empty (Figure 2) and thus raise suspicions in the viewer that there is nothing to be communicated. A way to avoid these suspicions is to fill up the plot with nondata figurations—what Tufte has termed 'chartjunk'. Figure 3 shows a plot of the labor productivity

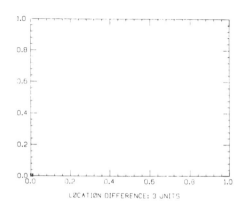

**Figure 2.** A low density graph (from Friedman and Rafsky 1981 (ddi = .5)).

**Figure 3.** A low density graph (© 1978, *The Washington Post*) with chart-junk to fill in the space (ddi = .2).

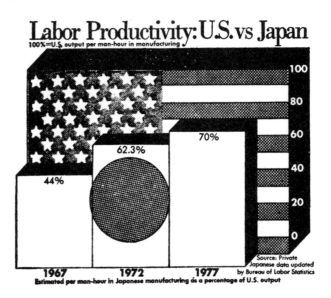

of Japan relative to that of the United States. It contains one number for each of three years. Obviously, a graph of such sparse information would have a lot of blank space, so filling the space hides the paucity of information from the reader.

A convenient measure of the extent to which this practice is in use is Tufte's 'data-ink ratio'. This measure is the ratio of the amount of ink used in graphing the data to the total amount of ink in the graph. The closer to zero this ratio gets, the worse the graph. The notion of the data-ink ratio brings us to the second principle of bad data display.

*Rule 2—Hide what data you do show (minimize the data-ink ratio)*

One can hide data in a variety of ways. One method that occurs with some regularity is hiding the data in the grid. The grid is useful for plotting the points, but only rarely afterwards. Thus to display data badly, use a fine grid and plot the points dimly (see Tufte 1983, pp 94–5 for one repeated version of this).

256

**Figure 4.** Hiding the data in the scale (from *SI3*).

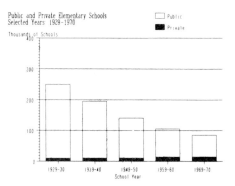

A second way to hide the data is in the scale. This corresponds to blowing up the scale (i.e., looking at the data from far away) so that any variation in the data is obscured by the magnitude of the scale. One can justify this practice by appealing to 'honesty requires that we start the scale at zero', or other sorts of sophistry.

In Figure 4 is a plot that (from SI3) effectively hides the growth of private schools in the scale. A redrawing of the number of private schools on a different scale conveys the growth that took place during the mid-1950's (Figure 5).

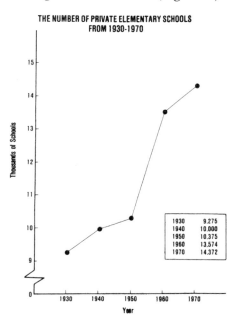

**Figure 5.** Expanding the scale and showing the data in Figure 4 (from *SI3*).

The relationship between this rise and *Brown vs. Topeka School Board* becomes an immediate question.

To conclude this section, we have seen that we can display data badly either by not including them (Rule 1) or by hiding them (Rule 2). We can measure the extent to which we are successful in excluding the data through the data density; we can sometimes convince viewers that we have included the data through the incorporation of chartjunk. Hiding the data can be done either by using an overabundance of chartjunk or by cleverly choosing the scale so that the data disappear. A measure of the success we have achieved in hiding the data is through the data-ink ratio.

### 3. Showing data accurately

The essence of a graphic display is that a set of numbers having both magnitudes and an order are represented by an appropriate visual metaphor—the magnitude and order of the metaphorical representation match the numbers. We can display data badly by ignoring or distorting this concept.

*Rule 3—Ignore the visual metaphor altogether*
If the data are ordered and if the visual metaphor has a natural order, a bad display will surely emerge if you shuffle the relationship. In Figure 6 note that the bar labeled

**Figure 6.** Ignoring the visual metaphor (© 1978, *The New York Times*).

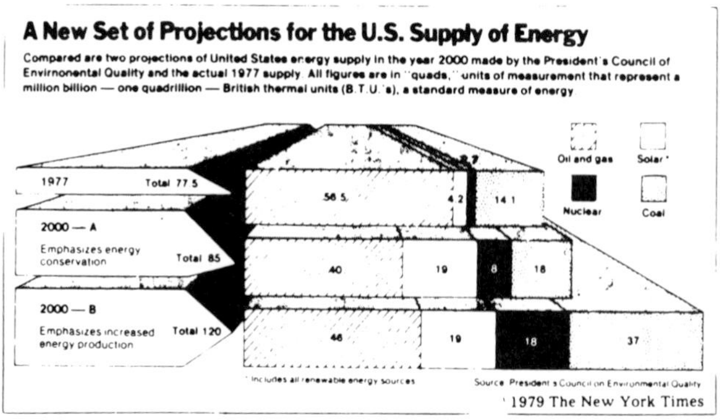

**Figure 7.** Reversing the metaphor in mid-graph while changing scales on both axes (© June 14, 1981, *The New York Times*).

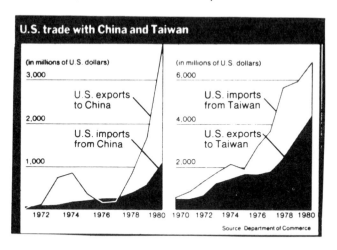

14.1 is longer than the bar labeled 18. Another method is to change the meaning of the metaphor in the middle of the plot. In Figure 7 the dark shading represents imports on one side and exports on the other. This is but one of the problems of this graph; more serious still is the change of scale. There is also a difference in the time scale, but that is minor. A common theme in Playfair's (1786) work was the difference between imports and exports. In Figure 8, a 200-year-old graph tells the story clearly. Two such plots would have illustrated the story surrounding this graph quite clearly.

*Rule 4—Only order matters*
One frequent trick is to use length as the visual metaphor when area is what is perceived. This was used quite effectively by *The Washington Post* in Figure 9. Note that this graph also has a low data density (.1), and its data-ink ratio is close to zero. We can also calculate Tufte's (1983) measure of perceptual distortion (PD) for this graph. The PD in this instance is the perceived change in the value of

**Figure 8.** A plot on the same topic done well two centuries earlier (from *Playfair* 1786).

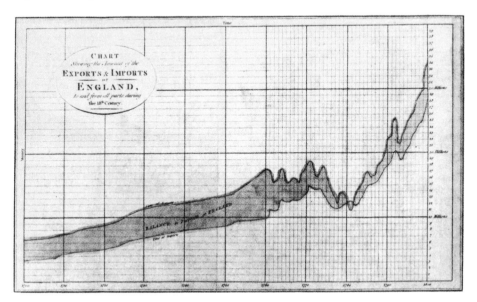

**Figure 9.** An example of how to goose up the effect by squaring the eyeball (© 1978, *The Washington Post*).

the dollar from Eisenhower to Carter divided by the actual change. I read and measure thus:

$$Actual: \frac{1.00 - .44}{.44} = 1.27 \qquad Measured: \frac{22.00 - 2.06}{2.06} = 9.68$$

PD = 9.68/1.27 = 7.62.

This distortion of over 700% is substantial but by no means a record.

A less distorted view of these data is provided in Figure 10. In addition, the spacing suggested by the presidential faces is made explicit on the time scale.

*Rule 5—Graph data out of context*
Often we can modify the perception of the graph (particularly for time series data) by choosing carefully the interval displayed. A precipitous drop can disappear if we choose a starting date just after the drop. Similarly, we can turn slight meanders into sharp changes by focusing on a single meander and expanding the scale. Often the choice of scale is arbitrary but can have profound effects on the perception of the display. Figure 11 shows a famous example in which President Reagan gives an out-of-context

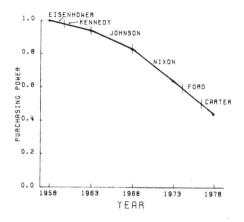

**Figure 10.** The data in Figure 9 as an unadorned line chart (from Wainer, 1980).

**Figure 11.** The White House showing neither scale nor context (© 1981, *The New York Times*).

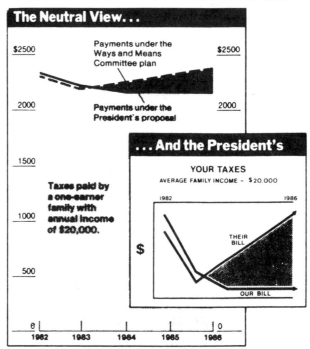

THE NEW YORK TIMES, SUNDAY, AUGUST 2, 1981

view of the effects of his tax cut. The *Times'* alternative provides the context for a deeper understanding. Simultaneously omitting the context as well as any quantitative scale is the key to the practice of Ordinal Graphics (see also Rule 4). Automatic rules do not always work, and wisdom is always required.

In Section 3 we discussed three rules for the accurate display of data. One can compromise accuracy by ignoring visual metaphors (Rule 3), by only paying attention to the order of the numbers and not their magnitude (Rule 4), or by showing data out of context (Rule 5). We advocated the use of Tufte's measure of perceptual distortion as a way of measuring the extent to which the accuracy of the data has been compromised by the display. One can think of modifications that would allow it to be applied in other situations, but we leave such expansion to other accounts.

### 4. Showing data clearly

In this section we discuss methods for badly displaying data that do not seem as serious as those described previously; that is, the data are displayed, and they might even be accurate in their portrayal. Yet subtle (and not so subtle) techniques can be used to effectively obscure the most meaningful or interesting aspects of the data. It is more difficult to provide objective measures of presentational clarity, but we rely on the reader to judge from the examples presented.

*Rule 6—Change scales in mid-axis*
This is a powerful technique that can make large differences look small and make exponential changes look linear.

In Figure 12 is a graph that supports the associated story

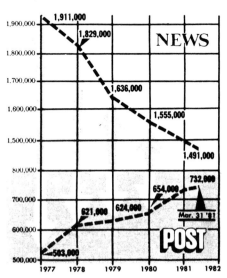

**Figure 12.** Changing scale in midaxis to make large differences small (© 1981, *New York Post*).

263

about the skyrocketing circulation of *The New York Post* compared to the plummeting *Daily News* circulation. The reason given is that New Yorkers 'trust' the *Post*. It takes a careful look to note the 700,000 jump that the scale makes between the two lines.

In Figure 13 is a plot of physicians' incomes over time. It appears to be linear, with a slight tapering off in recent years. A careful look at the scale shows that it starts out plotting every eight years and ends up plotting yearly. A more regular scale (in Figure 14) tells quite a different story.

*Rule 7—Emphasize the trivial (ignore the important)*
Sometimes the data that are to be displayed have one important aspect and others that are trivial. The graph can be made worse by emphasizing the trivial part. In Figure 15 we have a page from *SI3* that compares the income

**Figure 13.** Changing scale in mid-axis to make exponential growth linear (© *The Washington Post*).

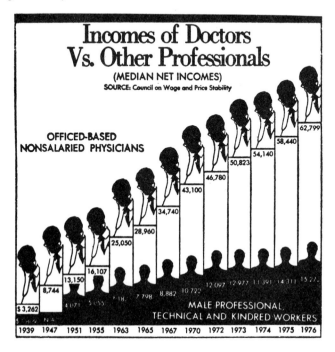

**Figure 14.** Data from Figure 13 redone with linear scale (from Wainer 1980).

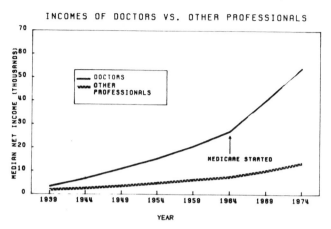

levels of men and women by educational levels. It reveals the not surprising result that better educated individuals are paid better than more poorly educated ones and that changes across time expressed in constant dollars are reasonably constant. The comparison of greatest interest and current concern, comparing salaries between sexes within education level, must be made clumsily by vertically transposing from one graph to another. It seems clear that Rule 7 must have been operating here, for it would have been easy to place the graphs side by side and allow the comparison of interest to be made more directly. Looking at the problem from a strictly data-analytic point of view, we note that there are two large main effects (education and sex) and a small time effect. This would have implied a plot that showed the large effects clearly and placed the smallish time trend into the background (Figure 16).

*Rule 8—Jiggle the baseline*
Making comparisons is always aided when the quantities being compared start from a common base. Thus we can always make the graph worse by starting from different bases. Such schemes as the hanging or suspended rootogram and the residual plot are meant to facilitate comparisons. In Figure 17 is a plot of U.S. imports of red

**Figure 15.** Emphasizing the trivial: Hiding the main effect of sex differences in income through the vertical placement of plots (from *S13*).

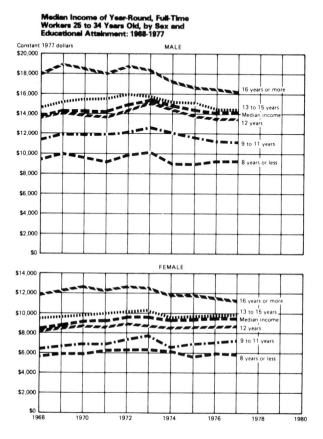

meat taken from the *Handbook of Agricultural Charts* published by the U.S. Department of Agriculture. Shading beneath each line is a convention that indicates summation, telling us that the amount of each kind of meat is added to the amounts below it. Because of the dominance of and the fluctuations in importation of beef and veal, it is hard to see what the changes are in the other kinds of meat—Is the importation of pork increasing? Decreasing? Staying constant? The only purpose for stacking is to indicate graphically the total summation. This is easily done through the addition of another line for

**Figure 16.** Figure 15 redone with the large main effects emphasized and the small one (time trends) suppressed.

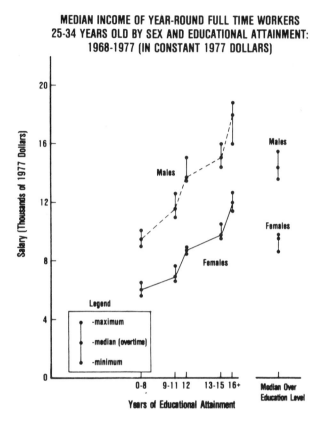

MEDIAN INCOME OF YEAR-ROUND FULL TIME WORKERS
25-34 YEARS OLD BY SEX AND EDUCATIONAL ATTAINMENT:
1968-1977 (IN CONSTANT 1977 DOLLARS)

TOTAL. Note that a TOTAL will always be clear and will never intersect the other lines on the plot. A version of these data is shown in Figure 18 with the separate amounts of each meat, as well as a summation line, shown clearly. Note how easily one can see the structure of import of each kind of meat now that the standard of comparison is a straight line (the time axis) and no longer the import amount of those meats with great volume.

*Rule 9—Austria first!*
Ordering graphs and tables alphabetically can obscure structure in the data that would have been obvious had

267

**Figure 17.** Jiggling the baseline makes comparisons more difficult (from *Handbook of Agricultural Charts*).

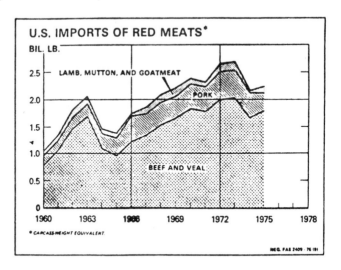

**Figure 18.** An alternative version of Figure 17 with a straight line used as the basis of comparison.

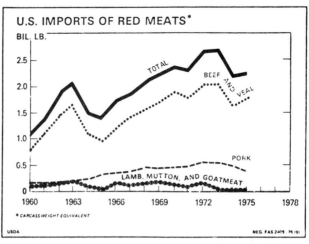

Source: <u>Handbook of Agricultural Charts</u>, U.S. Department of Agriculture, 1976, p. 93.
Chart Source: Original

268

the display been ordered by some aspect of the data. One can defend oneself against criticisms by pointing out that alphabetizing 'aids in finding entries of interest.' Of course, with lists of modest length such aids are unnecessary; with longer lists the indexing schemes common in 19th century statistical atlases provide easy lookup capability.

Figure 19 is another graph from *SI3* showing life expectancies, divided by sex, in 10 industrialized nations. The order of presentation is alphabetical (with the USSR positioned as Russia). The message we get is that there is little

**Figure 19.** Austria First! Obscuring the data structure by alphabetizing the plot (from *SI3*).

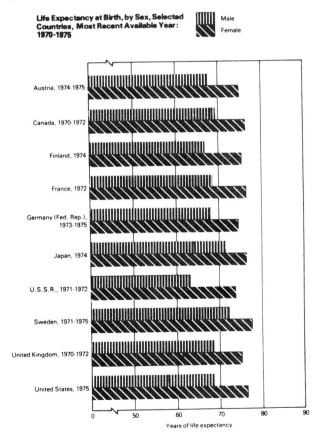

269

variation and that women live longer than men. Redone as a stem-and-leaf diagram (Figure 20 is simply a reordering of the data with spacing proportional to the numerical differences), the magnitude of the sex difference leaps out at us. We also note that the USSR is an outlier for men.

*Rule 10—Label (a) illegibly, (b) incompletely, (c) incorrectly, and (d) ambiguously*

There are many instances of labels that either do not tell the whole story, tell the wrong story, tell two or more stories, or are so small that one cannot figure out what story they are telling. One of my favorite examples of small labels is from *The New York Times* (August 1978), in which the article complains that fare cuts lower commission payments to travel agents. The graph (Figure 21) supports this view until one notices the tiny label indicating that the small bar showing the decline is for just the first half of 1978. This omits such heavy travel periods as Labor Day, Thanksgiving, Christmas, and so on, so that merely doubling the first-half data is probably not enough. Nevertheless, when this bar is doubled (Figure 22), we see that the agents are doing very well indeed compared to earlier years.

LIFE EXPECTANCY AT BIRTH, BY SEX,
MOST RECENT AVAILABLE YEAR

| WOMEN | YEARS | MEN |
|---|---|---|
| SWEDEN | 78 | |
| | 77 | |
| FRANCE, US, JAPAN, CANADA | 76 | |
| FINLAND, AUSTRIA, UK | 75 | |
| USSR, GERMANY | 74 | |
| | 73 | |
| | 72 | SWEDEN |
| | 71 | JAPAN |
| | 70 | |
| | 69 | CANADA, UK, US, FRANCE |
| | 68 | GERMANY, AUSTRIA |
| | 67 | FINLAND |
| | 66 | |
| | 65 | |
| | 64 | |
| | 63 | USSR |
| | 62 | |

**Figure 20.** Ordering and spacing the data from Figure 19 as a stem-and-leaf diagram provides insights previously difficult to extract (from *SI3*).

**Figure 21.** Mixing a changed metaphor with a tiny label reverses the meaning of the data (© 1978, *The New York Times*).

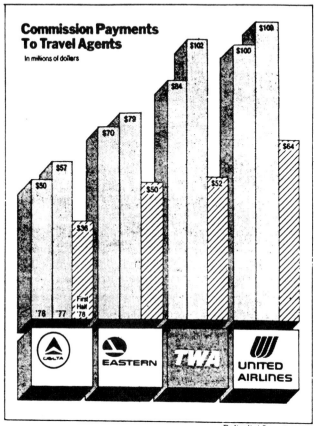

Complex web of discount fares and airlines' telephone delays are raising travel agents' overhead, offsetting revenue gains from higher volume.

*Rule 11—More is murkier: (a) more decimal places and (b) more dimensions*

We often see tables in which the number of decimal places presented is far beyond the number that can be perceived by a reader. They are also commonly presented to show more accuracy than is justified. A display can be made clearer by presenting less. In Table 1 is a section of a table from Dhariyal and Dudewicz's (1981) JASA paper. The table entries are presented to five decimal places! In Table

271

**Figure 22.** Figure 21 redrawn with 1978 data placed on a comparable basis (from Wainer 1980).

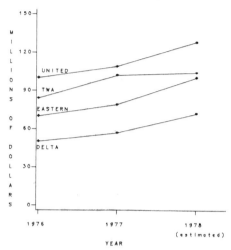

*Commission Payments to Travel Agents*

2 is a heavily rounded version that shows what the authors intended clearly. It also shows that the various columns might have a substantial redundancy in them (the maximum expected gain with $b/c = 10$ is about 1/10th that of $b/c = 100$ and 1/100th that of $b/c = 1,000$). If they do, the entire table could have been reduced substantially.

Just as increasing the number of decimal places can make a table harder to understand, so can increasing the

### Table 1. Optimal Selection From a Finite Sequence With Sampling Cost

| | $b/c = 10.0$ | | 100.0 | | 1,000.0 | |
|---|---|---|---|---|---|---|
| N | $r^*$ | $(G_N(r^*) - a)/c$ | $r^*$ | $(G_N(r^*) - a)/c$ | $r^*$ | $(G_N(r^*) - a)/c$ |
| 3 | 2 | .20000 | 2 | 2.22500 | 2 | 22.47499 |
| 4 | 2 | .26333 | 2 | 2.88833 | 2 | 29.13832 |
| 5 | 2 | .32333 | 3 | 3.54167 | 3 | 35.79166 |
| 6 | 3 | .38267 | 3 | 4.23767 | 3 | 42.78764 |
| 7 | 3 | .44600 | 3 | 4.90100 | 3 | 49.45097 |
| 8 | 3 | .50743 | 4 | 5.57650 | 4 | 56.33005 |
| 9 | 3 | .56743 | 4 | 6.26025 | 4 | 63.20129 |
| 10 | 4 | .62948 | 4 | 6.92358 | 4 | 69.86462 |

NOTE: $g(Xs + r - 1) = bR(Xs + r - 1) + a$, if $S = s$, and $g(Xs + r - 1) = 0$, otherwise.
Source: Dhariyal and Dudewicz (1981).

272

*Table 2. Optimal Selection From a Finite Sequence*
*With Sampling Cost (revised)*

| N | b/c = 10 | | b/c = 100 | | b/c = 1,000 | |
|---|---|---|---|---|---|---|
| | r* | G | r* | G | r* | G |
| 3 | 2 | .2 | 2 | 2.2 | 2 | 22 |
| 4 | 2 | .3 | 2 | 2.9 | 2 | 29 |
| 5 | 2 | .3 | 3 | 3.5 | 3 | 36 |
| 6 | 3 | .4 | 3 | 4.2 | 3 | 43 |
| 7 | 3 | .4 | 3 | 4.9 | 3 | 49 |
| 8 | 3 | .5 | 4 | 5.6 | 4 | 56 |
| 9 | 3 | .6 | 4 | 6.3 | 4 | 63 |
| 10 | 4 | .6 | 4 | 6.9 | 4 | 70 |

NOTE: $g(Xs + r - 1) = bR(Xs + r - 1) + a$, if $S = s$, and $g(Xs + r - 1) = 0$, otherwise.

number of dimensions make a graph more confusing. We have already seen how extra dimensions can cause ambiguity (Is it length or area or volume?). In addition, human perception of areas is inconsistent. Just what is confusing and what is not is sometimes only a conjecture, yet a hint that a particular configuration will be confusing is obtained if the display confused the grapher. Shown in Figure 23 is a plot of per share earnings and dividends over a six-year period. We note (with some amusement) that 1975 is the side of a bar—the third dimension of this bar (rectangular parallelopiped?) chart has confused the artist! I suspect that 1975 is really what is labeled 1976, and the unlabeled bar at the end is probably 1977. A simple line chart with this interpretation is shown in Figure 24.

In Section 4 we illustrate six more rules for displaying data badly. These rules fall broadly under the heading of how to obscure the data. The techniques mentioned were to change the scale in mid-axis, emphasize the trivial, jiggle the baseline, order the chart by a characteristic unrelated to the data, label poorly, and include more dimensions or decimal places than are justified or needed. These methods will work separately or in combination with others to produce graphs and tables of little use. Their common effect will usually be to leave the reader uninformed about the points of interest in the data,

273

**Figure 23.** An extra dimension confuses even the grapher (© 1979, *The Washington Post*).

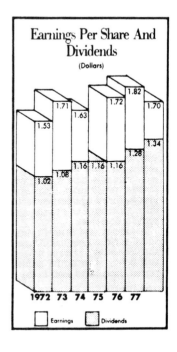

although sometimes they will misinform us; the physicians' income plot in Figure 13 is a prime example of misinformation.

Finally, the availability of color usually means that there are additional parameters that can be misused. The U.S. Census' two-variable color map is a wonderful example of how using color in a graph can seduce us into thinking that we are communicating more than we are (see Fienberg 1979; Wainer and Francolini 1980; Wainer 1981). This leads us to the last rule.

*Rule 12—If it has been done well in the past, think of another way to do it*
The two-variable color map was done rather well by Mayr (1874), 100 years before the U.S. Census version. He used bars of varying width and frequency to accomplish gracefully what the U.S. Census used varying saturations to do clumsily.

A particularly enlightening experience is to look carefully through the six books of graphs that William

**Figure 24.** Data from Figure 23 redrawn simply (from Wainer 1980).

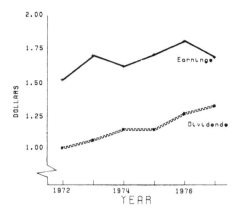

Playfair published during the period 1786 – 1822. One discovers clear, accurate, and data-laden graphs containing many ideas that are useful and too rarely applied today. In the course of preparing this article, I spent many hours looking at a variety of attempts to display data. Some of the horrors that I have presented were the fruits of that search. In addition, jewels sometimes emerged. I saved the best for last, and will conclude with one of those jewels—my nominee for the title of 'World's Champion Graph'. It was produced by Minard in 1861 and portrays the devastating losses suffered by the French army during the course of Napoleon's ill-fated Russian campaign of 1812. This graph (originally in color) appears in Figure 25 and is reproduced from Tufte's book (1983, p. 40). His narrative follows.

Beginning at the left on the Polish-Russian border near the Nieman River, the thick band shows the size of the army (422,000 men) as it invaded Russia in June 1812. The width of the band indicates the size of the army at each place on the map. In September, the army reached Moscow, which was by then sacked and deserted, with 100,000 men. The path of Napoleon's retreat from Moscow is depicted by the darker, lower band, which is linked to a temperature scale and dates at the bottom of the chart. It was a bitterly cold winter, and many froze on the march out of Russia. As the graphic shows, the crossing of the Berezina River was a disaster, and the army finally struggled back to Poland with only 10,000 men remaining. Also shown are the movements of auxiliary troops, as they sought to protect the rear and flank of the advancing army. Minard's graphic tells a rich, coherent story with its multivariate data, far more enlightening than just a single number bouncing along over time. *Six* variables are plotted: the size of the army, its location on a two-dimensional surface, direction of

275

**Figure 25.** Minard's (1861) graph of the French Army's ill-fated foray into Russia – A candidate for the title of 'World's Champion Graph' (see Tufte 1983 for a superb reproduction of this in its original color—p 176).

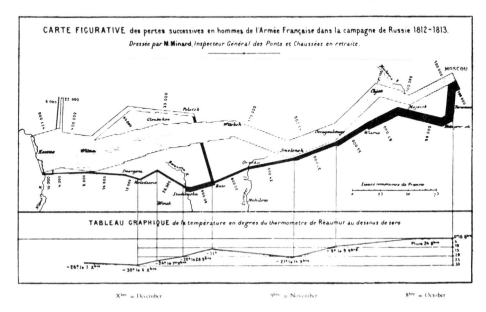

the army's movement, and temperature on various dates during the retreat from Moscow.

It may well be the best statistical graphic ever drawn.

## 5. Summing up

Although the tone of this presentation tended to be light and pointed in the wrong direction, the aim is serious. There are many paths that one can follow that will cause deteriorating quality of our data displays; the 12 rules that we described were only the beginning. Nevertheless, they point clearly toward an outlook that provides many hints for good display. The measures of display described are interlocking. The data density cannot be high if the graph is cluttered with chartjunk; the data-ink ratio grows with the amount of data displayed, perceptual distortion manifests itself most frequently when additional dimensions or worthless metaphors are included. Thus, the rules for good display are quite simple. Examine the data carefully enough to know what they have to say, and then let them say it with a minimum of adornment. Do this while following reasonable regularity practices in the depiction of scale, and label clearly and fully. Last, and perhaps most important, spend some time looking at the work of the masters of the craft. An hour spent with Playfair or Minard will not only benefit your graphical expertise but will also be enjoyable. Tukey (1977) offers 236 graphs and little chartjunk. The work of Francis Walker (1894) concerning statistical maps is clear and concise, and it is truly a mystery that their current counterparts do not make better use of the schema developed a century and more ago.

## References

BERTIN, J. (1973), *Semiologie Graphique* (2nd ed.), The Hague: Mouton-Gautier.

—— (1977), *La Graphique et le Traitement Graphique de l'Information*, France: Flammarion.

—— (1981), *Graphics and the Graphical Analysis of Data*, translation, W. Berg, tech. ed., H. Wainer, Berlin: DeGruyter.

COX, D.R. (1978), 'Some Remarks on the Role in Statistics of Graphical Methods,' *Applied Statistics*, 27, 4–9.

DHARIYAL, I.D., and DUDEWICZ, E.J. (1981), 'Optimal Selection From a Finite Sequence With Sampling Cost.' *Journal of the American Statistical Association*, 76, 952–959.

EHRENBERG, A.S.C. (1977), 'Rudiments of Numeracy,' *Journal of the Royal Statistical Society*, Ser. A, 140, 277–297.

FIENBERG, S.E. (1979), Graphical Methods in Statistics, *The American Statistician*, 33, 165–178.

FRIEDMAN, J.H., and RAFSKY, L.C. (1981), 'Graphics for the Multivariate Two-Sample Problem,' *Journal of the American Statistical Association*, 76, 277–287.

JOINT COMMITTEE ON STANDARDS FOR GRAPHIC PRESENTATION, PRELIMINARY REPORT (1915), *Journal of the American Statistical Association*, 14, 790–797.

MACDONALD-ROSS, M. (1977), 'How Numbers Are Shown: A Review of Research on the Presentation of Quantitative Data in Texts,' *Audiovisual Communications Review*, 25, 359–409.

MAYR, G. von (1874), 'Gutachen Uber die Anwendung der Graphischen und Geographischen,' *Method in der Statistik*, Munich.

MINARD, C.J. (1845–1869), *Tableaus Graphiques et Cartes Figuratives de M. Minard*, Bibliothéque de l'Ecole Nationale des Ponts et Chaussées, Paris.

How to display data badly

PLAYFAIR, W. (1786), *The Commercial and Political Atlas*, London: Corry.

SCHMID, C.F. (1954), *Handbook of Graphic Presentation*, New York: Ronald Press.

SCHMID, C.F., and SCHMID, S.E. (1979), *Handbook of Graphic Presentation* (2nd ed.), New York: John Wiley.

TUFTE, E.R. (1977), 'Improving Data Display,' University of Chicago, Dept. of Statistics.

—— (1983), *The Visual Display of Quantitative Information*, Cheshire, Conn.: Graphics Press.

TUKEY, J.W. (1977), *Exploratory Data Analysis*, Reading, Mass: Addison-Wesley.

WAINER, H. (1980), 'Making Newspaper Graphs Fit to Print,' in *Processing of Visible Language*, Vol. 2, eds. H. Bouma, P.A. Kolers, and M.E. Wrolsted, New York: Plenum, 125–142.

——, 'Reply' to Meyer and Abt (1981), *The American Statistician*, 57.

—— (1983), 'How Are We Doing? A Review of Social Indicators III,' *Journal of the American Statistical Association*, 78, 492–496.

WAINER, H., and FRANCOLINI, C. (1980), 'An Empirical Inquiry Concerning Human Understanding of Two-Variable Color Maps,' *The American Statistician*, 34, 81–93.

WAINER, H., and THISSEN, D. (1981), 'Graphical Data Analysis,' *Annual Review of Psychology*, 32, 191–214.

WALKER, F.A. (1894), *Statistical Atlas of the United States Based on the Results of the Ninth Census*, Washington, D.C.: U.S. Bureau of the Census.

[TUFTE, E.R. (1990), *Envisioning Information*, Cheshire, Conn.: Graphics Press.]

# Confusion, Frustration

The only way for a new theory to become accepted is for adherents of the old theories to die.

*Max Planck*

Deep in the human unconscious is a pervasive need for a logical universe that makes sense. But the real universe is always one step beyond logic.

*Frank Herbert*

# The certainty of uncertainty

RBP from Adrian Berry *From Apes to Astronauts*, © The Daily Telegraph plc, 1980.

In the celebrations of Albert Einstein's centenary year, little has been said of the late Werner Heisenberg, the only man to have contradicted Einstein and be proved right, and who laid the cornerstone for a new science as important as relativity itself.

'Heisenberg may have been here', says a joke which sums up quantum mechanics, which even today seems as mysterious as the personality of its founder. Heisenberg's famous principle of uncertainty states that certain information is for ever unobtainable. This is not due to any deficiency of information-gathering, but rather to the nature of knowledge itself. Consider this question: *Could an infinitely powerful computer make an infinitely accurate prediction?*

Since an infinitely powerful computer cannot exist, the question might at first sight seem merely philosophical. But the question is whether only the absence of supremely sophisticated instruments hides from us the totality of knowledge.

The Marquis of Laplace, in the early 19th century, suggested that God must be infinitely well-informed, even if man was not. A supreme intelligence, he said, 'which knew at a given instant all of the forces by which nature is animated, and the relative positions of all the objects, could include in one formula the movements of the massive objects in the universe and those of the lightest atom. Nothing would be uncertain to it. The future, as the past, would be present to its eyes.'

Werner Heisenberg, in his great flash of insight of 1927, afterwards confirmed by experiment, showed that Laplace was wrong and that the answer to the question was 'no'. Not even God could possess non-existent information.

His great uncertainty principle showed that electrons, the particles that orbit the nuclei of atoms, can never be located with absolute accuracy. The more precisely we define their positions, the less we know of their speeds. And the more we know of their speeds, the less we know of their positions.

This is not just a question of the information being un-

280

obtainable because of inadequate instruments. The complete information does not exist, since the behaviour of the electron is itself altered by the presence of the observer. In short, the popular expression 'God knows!' may at times be hopelessly optimistic.

Heisenberg's discovery was a shock to many people, since the supposed principle of cause and effect had been an anchor of science since the days of the ancient Greeks. Einstein himself was profoundly disturbed by it, protesting that 'God does not play the dice.' 'You can't tell God what to do', answered his colleague Neils Bohr. This year the cosmologist Stephen Hawking took the paradox even further: 'Not only does God play at dice; He sometimes throws the dice where they can't be seen.'

The uncertainty principle, so far as we can tell, applies to the micro-system of electrons rather than the larger universe of planets, stars and galaxies.

Yet this prohibition is not absolute. The moon, according to Issac Newton's first law, will continue to orbit the earth in some trajectory or other until the end of time, unless acted on by a force. The orbit will change as the earth and moon exchange tidal drag, but it will be an orbit whose path can be predicted with almost absolute accuracy.

*Almost* absolute. That is the point. The moon just might, for no apparent reason, cease its orbit and crash into the earth. The odds against this happening during a period equal to the lifespan of the universe are so remote as to involve a number almost beyond possibility of calculation, but they are not zero.

It is amusing to speculate whether the uncertainty principle could apply also to the human personality. Like the electron of which we seek in vain to know two things, its position and its speed, the human personality has two distinct sides, the intellectual and the emotional. Our information about one side cancels out our information about the other. We may know a great deal about a particular person's intellect, or about their emotions. We may know everything either about the one or the other, but we cannot, it seems, know everything about both.

The reason appears plain. When people are thinking sufficiently deeply, their emotions are hidden. When in a strongly emotional state, their capacity for rational thought appears correspondingly to have diminished.

Since a person's mental state is usually the result of external events or other people's behaviour, we can say that the personality, like the electron, is itself altered by the presence of the observer.

Could this uncertainty explain our fascination for the human mind? Could the unpredictability of human nature explain our love of art, because certain things are both unknown and unknowable? I offer the idea for what it is worth.

---

# Technology succumbs to success

CORNING
Consumer Products Division
Corning Glass Works
Corning, New York  14830

December 8, 1977

Mr Robert L Weber
625 West Ridge Avenue
State College, PA 16801

Dear Mr Weber:

Thank you for writing to us about your CENTURA® dinnerware.

The decision has been made to discontinue this line of dinnerware because we no longer have the plant capacity to manufacture it in sufficient quantities to meet consumer needs. ...

We share your disappointment and hope you can understand why it was necessary to take this action.

Sincerely,

(Mrs) Emma Kuehnle
Consumer Service Department


[When subsequently questioned, the manufacturers explained
that the letter had been badly worded and that what was meant
was that the line of dinnerware was discontinued due to lack of
demand.]

---

# The science of why nothing works the way it's supposed to

*John Gall†*

[Despite Parkinson's Law and Peter's Principle, organizations
are still operating most peculiarly. The new science of Systemantics explains it all.]

Appeared in The
New York Times
Magazine on 26
December 1976, p
10. © John Gall. A colleague of the author (let us call him Jones), considered himself fortunate to be employed as research
associate at a small state home for retarded children. In
this humble, even despised, position he was too low on
the Civil Service scale to merit the attention of
administrators, and he was therefore left alone to tinker
with ideas in his chosen field. He was happily pursuing
his own research interest when, following Presidential
interest and national publicity, mental retardation suddenly became a fashionable subject. Jones received an
urgent invitation to join an ambitious Federally funded

†John Gall is a pediatrician who practices in Ann Arbor, Mich. This article
is adapted from *Systemantics* 1987 (Quadrangle/The New York Times Publishing
Co).

project for a systematic attack upon the problem of mental retardation.

Thinking that his new job, with its ample funds and facilities, would advance both his research efforts and his career, Jones joined. Within three months, his own research had come to a halt, and within a year he was completely unable to speak or think intelligently in the field of mental retardation. He had, in fact, become *retarded* relative to his previous condition.

Looking about him to discover, if possible, what had happened, Jones found that a dozen other skilled professionals who made up the staff of the project had experienced the same catastrophe. What had gone wrong? Jones, knowing of the author's interest in systems operation, turned to him for advice.

We first of all reviewed C Northcote Parkinson's classic essay on institutional paralysis, hoping to find enlightenment there.

Was the disease a fulminating case of injelititis (Parkinson's term for the condition that results when an organization falls into the hands of administrators consumed with both incompetence and jealousy)? Obviously not. The professional staff of the institute was competent, dedicated and hard-working. Was it an example of Laurence J Peter's famous Principle, in which members had ascended the hierarchy until they had reached jobs for which they were not fitted? No. This was a new organization, and few had been promoted.

Slowly, it began to dawn on us that men do not yet understand the basic laws governing the behavior of complex organizations. Faced with this realization, and moved by the dramatic and touching crisis that had overtaken his colleague, the author resolved to redouble his researches into the causes of organizational ineptitude and systems malfunction. Little did he suspect, at that moment, that those studies would lead to the new science of Systemantics.

284

## I. Historical overview

All over the world, in great metropolitan centers as well as in the remotest rural backwater, in sophisticated electronics laboratories and in dingy clerical offices, people are struggling with a problem:

**Things aren't working very well.**

This, of course, is nothing new. People have been discouraged about things in general many times in the past. A good deal of discouragement prevailed during the Dark Ages, and morale was rather low in the later Middle Ages, too. The Industrial Revolution brought with it depressing times, and the Victorian Era was felt by many to be particularly gloomy. At all times there have been people who felt that *Things weren't working out very well.* This observation has gradually won recognition as an unavoidable, if inconvenient, fact of life, an inseparable component of the human condition. Because of its central role in all that follows (being the fundamental observation upon which all further research into systems has been based) it is known as the PRIMAL SCENARIO. We give it here in full:

**Things (things generally/all things/the whole works) are indeed not working very well. In fact, they never did.**

In formal Systemantics terminology, it may be stated concisely in axiomatic form:

**Systems in general work poorly or not at all.**

The semilegendary, almost anonymous Murphy (floreat circa 1940?) chose to disguise his genius by stating a fundamental systems theorem in commonplace, almost pedestrian terminology. This law, known to schoolboys the world over as *Jellybread always falls jelly-side down*, is here restated in Murphy's own words, as it appears on the walls of most of the world's scientific laboratories: *If anything can go wrong, it will.*

Let us ask what lies beneath the surface of that melancholy but universal observation. Just what is the characteristic feature of things not working out? To ask the question is to see the answer, almost immediately. It is the element of paradox.

ITEM: Insecticides, introduced to control disease and improve crop yields, turn up in the fat pads of auks in the antipodes and in the eggs of osprey in the Orkneys, resulting in incalculable ecological damage.

ITEM: The Aswan Dam, built at enormous expense to improve the lot of the Egyptian peasant, is said to have caused the Nile to deposit its fertilizing sediment in Lake Nasser, where it is unavailable. Egyptian fields must now be artificially fertilized. Fertilizer plants have been built to meet the new need. The plants require enormous amounts of electricity. So a lot of the dam's electrical output goes to supply the increased need for electricity that was created by the building of the dam.

ITEM: Many backward nations, whose great need is food to feed their people, sell their crops and bankrupt themselves to buy—not food—but advanced military hardware for the purpose of defending themselves against their equally backward neighbors, who are doing the same thing.

ITEM: It is reported that when Jimmy Carter was Governor of Georgia, he decided to save on heating bills by turning down the executive mansion thermostat to 65 degrees. But he was astonished to find that in the first month of the reduced setting, the heating bill had skyrocketed. An aide explained that the mansion was equipped with a system that maintained constant temperature with a combination of heating and air-conditioning. If you turned down the temperature within a certain range, the heat stayed on, but the air-conditioners started up to keep the place cool. Carter wanted to replace this wasteful system, but found that it would cost more to do so than to pay higher heating bills.

286

## II. Things are not what they seem

There is a man in our neighborhood who is building a boat in his backyard. He knows very little of boatbuilding and still less of sailing or navigation. He works from plans drawn up by himself. Nevertheless, he is demonstrably building a boat and can be called, in some real sense, a boatbuilder.

Now if you go down to Hampton Roads or any other shipyard and look around for a shipbuilder, you will be disappointed. You will find—in abundance—welders, carpenters, foremen, engineers and many other specialists, but no shipbuiders. True, the company executives may call themselves shipbuilders, but if you observe them at their work, you will see that it really consists of writing contracts, planning budgets and other administrative activities. Clearly, they are not in any concrete sense building ships. In cold fact, a SYSTEM is building ships, and the SYSTEM is the shipbuilder. We conclude from the above that:

> **People in systems do not do what the system says they are doing.**

This paradox was clearly recognized and described in detail by the 19th century team of empirical sociologists, Gilbert and Sullivan, when they wrote, 'But that kind of ship so suited me/that now I am the Ruler of the Queen's Navee.' The 'ship' referred to, the reader will recall, was the Admiral's legal *partner*ship.

In general, the larger and more complex the system, the less the resemblance between the true function and the name it bears. For brevity, we shall refer to this paradox as FUNCTIONARY'S FALSITY.

The OPERATIONAL FALLACY is merely the systems-analogue of FUNCTIONARY'S FALSITY. Just as *People in systems do not do what the system says they are doing*, so also:

> **The system itself does not do what it says it is doing.**

The reason we think the auto industry is meeting our

needs is that we have almost completely forgotten what we originally wanted, namely, a means of going from one place to another that would be cheap, easy, convenient, safe and fast. We have been brainwashed into thinking that the Detroit product meets these requirements.

**If Detroit makes it, it must be an automobile.**

### III. Ultimate models

Now, so long as one is content merely to make the observation that a particular system isn't working well, or isn't doing what was expected, one is really only at the level of insight summarized in the PRIMAL SCENARIO. The crucial step forward in logic is a small one , but it is of critical importance for progress in Systemantics-thinking. There is a world of difference, psychologically speaking, between the passive observation that *Things don't work out very well,* and the active, penetrating insight that:

**Complex systems exhibit unexpected behavior.**

The operation of this principle is perhaps most clearly displayed in the realm of CLIMAX DESIGN; e.g., in the construction of the largest and most complex examples of man-made systems, whether buildings, ships and planes or organizations. The ultimate model (the largest, fastest, tallest, etc.) often, if not invariably, exhibits behavior so unexpected as to verge on the uncanny. The behavior is often an unsuspected *way of failing.* Let us review only two examples of CLIMAX DESIGN: The largest building in the world, the space vehicle preparation shed at Cape Kennedy, *generates its own weather, including clouds and rain.* Designed to protect space rockets from the elements, it pelts them with storms of its own. The Queen Elizabeth 2, greatest ocean liner ever built, has three separate sets of boilers for safety, reliability and speed. Yet on a recent cruise, in fine weather and a calm sea, all three sets of boilers *failed simultaneously.*

## IV. Fractions of truth

Here we present the FUNDAMENTAL LAW OF ADMINISTRATIVE WORKINGS (F.L.A.W.):

**Things are what they are reported to be.**

This axiom has been stated in various ways, all to the same effect. Standard formulations include the following:

**The real world is what is reported to the system.**

**If it isn't official, it hasn't happened.**

The net effect of this law is to insure that people in systems are never dealing with the real world that the rest of us have to live in but with a filtered, distorted and censored version which is all that can get past the sensory organs of the system itself.

**To those within a system, the outside reality tends to pale and disappear.**

This effect has been studied in some detail by a small group of dedicated Systemanticists. In an effort to introduce quantitative methodology into this important area of research, they have paid particular attention to the amount of information that reaches, or fails to reach, the attention of the relevant administrative officer. The crucial variable, they have found, is the fraction:

**Ro/Rs**

where Ro equals the amount of reality that fails to reach the relevant administrative officer and Rs equals the total amount of reality presented to the system. The fraction Ro/Rs varies from zero (full awareness of outside reality) to unity (no reality getting through). It is known, of course, as the COEFFICIENT OF FICTION.

A high C.F. has particular implications for the relationship between the system and an individual person (represented by the lower-case letter *i*). We state the relationship as follows:

**The bigger the system, the narrower and more specialized the interface with individuals.**

In very large systems, the relationship is not with the individual at all but with his Social Security number, his driver's license, or some other paper phantom. In systems of medium size, some residual awareness of the individual may still persist. A hopeful indication was recently observed by the author in a medium-sized hospital. Taped to the wall of the nurses' station, just above the Vital Signs Remote Sensing Console, which enables the nurses to record whether the patient is breathing and even to take his pulse without actually going down the hall to see him, was the following handlettered reminder.

**The chart is not the patient.**

Unfortunately, this slogan, with its humanistic implications, turned out to be misleading. The nurses were neither attending the patients nor making notations in the charts. They were in the hospital auditorium, taking a course in interdisciplinary function. (Interdisciplinary function may be defined as the art of correlating one's own professional activities more and more with those of other professionals while actually doing less and less.)

## V. The entrapment of Trillium: a fable

Every student of science is required at some point or other in his career to learn Le Chatelier's Principle. Briefly, this law states that any natural process, whether physical or chemical, tends to set up conditions opposing the further operation of the process. No one who has had any experience with the operation of large systems can fail to appreciate its force, especially when it is stated in cogent Systemantics terminology, as follows:

**Systems get in the way**

or, alternatively:

**The system always kicks back.**

Let us take as an example the case of Lionel Trillium, a young assistant professor in the department of botany at Hollyoak College.

Trillium's department head, Baneberry, has for some years now failed to initiate new and interesting hypotheses about the behavior of the slime molds, his chosen area of specialization. Paralleling this decline of scientific productivity, he has exhibited increasing interest in improving the 'efficiency' of his department. He fires off a memo to the staff of his department requiring them to submit to him, in triplicate, by Monday next, statements of their Goals and Objectives.

This demand catches Trillium at a bad time. His studies of angiosperms are at a critical point. Nevertheless, he must take time out to consider his Goals and Objectives, as the wording of the memo leaves little doubt of the consequences of failure to comply.

Now Trillium really does have some personal goals of his own that his study of botany can advance. In actual fact, he entered that field because of unanswered questions having to do with the origin of babies. His boyhood curiosity was never satisfied, and it has become fixed on the mechanics of reproductive processes in living creatures. But to study reproduction directly, in animals, creates too much anxiety. Therefore, he has chosen the flowering plants, whose blatant sexuality is safely isolated from our own.

But now his chief is demanding Goals and Objectives. This is both disturbing and threatening. Trillium doesn't want to think about his real goals and objectives; indeed, they are unknown to his conscious mind. He only knows he likes botany.

But he can't reply in one line, 'I like botany and want to keep on studying it.' No, indeed! What is expected is a good deal more formal, more organized than that. It should fill at least three typewritten sheets, single-spaced, and should list Objectives and Sub-objectives in order of priority. Trillium goes into a depression just thinking about it.

Furthermore, he cannot afford to state his true goals. He must at all costs avoid giving the impression of an ineffective putterer or dilettante.

Trillium puts it off as long as possible. Finally, he gives up his research, stays home three days, and writes the damned thing.

But now he is committed in writing to a program. Because he has clearly stated his Goals and Objectives, it is now possible to deduce with rigorous logic how he should spend his waking and working hours in order to achieve them most efficiently. No more pottering around pursuing spontaneous impulses and temporary enthusiasms! No more happy hours in the departmental greenhouse! Just as a straight line is the shortest distance between two points, so an efficient worker will move from Sub-objective A to Sub-objective B in logical pursuit of Objective K, which leads in turn toward the Overall Goal. Trillium can be graded, not only on his *achievements* for the year, but also on the *efficiency* with which he moves toward each objective. He has become *administratively encircled*.

Baneberry, meanwhile, has been powerfully reinforced in his new-found function as judge of other botanists. His own botanical career may be wilting, but he has found a new career in assessing the strengths and weaknesses of his colleagues—especially their weaknesses.

## VI. Only the system knows

In accordance with our practice of moving from the simpler and more easily understandable to the more profound and impalpable, we present the following FUNCTIONAL INDETERMINACY THEOREM (F.I.T.):

> In complex systems, malfunction and even total nonfunction
> may not be detectable for long periods, if ever.

This theorem often elicits surprise when first propounded. However, illustrative examples abound, especially from history. For example, it would seem reasonable to suppose that absolute monarchies, Oriental despotisms and other governments in which all power is concentrated in the will of one man would require as a minimum for the adequate functioning of those governments that the will of the despot be intact. Nevertheless, the list of absolute monarchs who were hopelessly incompetent, even insane, is suprisingly long. They ruled with utter caprice, not to say whimsically, for decades on end, and the net result to their countries was— undetectably different from the rule of the wisest kings. (Students wishing to investigate this fascinating topic in more detail are advised to study the lives of Henry VIII, George III, certain emperors of Japan, the Czars of Russia, the Sultans of Turkey, etc., etc. Readers may also wish to review the performances of present-day heads of state.)

But, however difficult it may be to know what a system is doing, or even whether it is doing anything, we can still be sure of the validity of the Newtonian Law of Inertia as applied to systems:

> A system that performs a certain function or operates in a
> certain way will continue to operate in that way regardless of
> the need or of changed conditions.

In accordance with this principle, the Selective Service System continued to register young men for the draft long after the draft had ended.

## VII. How safe is fail-safe?

In the early days of the development of electronic computers, engineers were startled to observe that the probability of malfunction was proportional to the fourth power of the number of vacuum tubes. That observation led them to a preoccupation with component reliability that has culminated in modern transistorized, solid-state computers. But, impressive as that development may be, it is, from the standpoint of Systemantics, merely a neurotic digression from the straight and narrow path. Improvement in component reliability merely postpones the day of reckoning. As the system grows in size and complexity, it gradually but inevitably outgrows its component specifications. Parts (whether human or electronic) begin to fail. The important point is:

> **Any large system is going to be operating most of the time in failure mode.**

What the system is supposed to be doing when everything is working well is really beside the point, because that happy state is never achieved in real life. The truly pertinent question is: How does it work when its components aren't working well? How does it fail? How well does it function in failure mode?

Our basic approach is indicated in the Fundamental Failure Theorem (F.F.T.):

> **A system can fail in an infinite number of ways.**

An extreme example is the Government of Haiti, which consists entirely of departments that do not function. Dozens of national and international aid agencies, frustrated by the inability of the Haitian Government to cope with outside assistance, have sent emergency representatives to Haiti, to teach the Government officials *how to fill out requests for aid*.

The Fundamental Failure Theorem implies the Fail-Safe Theorem, which we believe to be of special immediacy for everyone concerned with the fate of Homo sapiens. We give it in all its austere and forbidding simplicity:

> **When a fail-safe system fails, it fails by failing to fail-safe.**

294

## VIII. Humble thanks

Although many of the world's frustrations are rooted in the malfunctions of complex systems, it is important to remember, finally, that *Some complex systems actually function*. This statement is not an axiom. It is an observation of a natural phenomenon. The obverse of the PRIMAL SCENARIO, it is not a truism, nor is there anything in modern philosophy that requires it to be true. We accept it here as a *given*, and offer humble thanks. The correct attitude of thankfulness leads naturally to a final law of Systemantics:

**If a system is working, leave it alone.**

---

# New fibre runs into trouble

### Land-use excessive, waste disposal impossible

RBP of the author, Trevor A Kletz.

The proposals by ICI and other chemical companies to invest millions of pounds in the production of a new fibre are already meeting widespread opposition. While nylon and polyester and the other fibres which have been in use since the dawn of civilization are manufactured in conventional plants, the new fibre, known as WOOL (Wildlife Origin Oily Ligament) will be grown on the backs of a specially developed breed of *Ovis musimon*.

Opposition to the new fibre centres on the extensive areas of land required for its production. While a million pounds per year of nylon or polyester can be produced in a fraction of an acre, the same quantity of WOOL will require at least 25 000 acres of good land or a larger area of hill land. This land will no longer be available for growing crops or for recreation.

For once the National Farmers' Union and the Ramblers' Association have combined to oppose development and a public enquiry will be necessary.

The RSPCA has protested at the 'industrialization' of animals and has asked what will happen if they break loose from their enclosures. Although *Ovis musimon* is docile, all animal species, even man, produce occasional agressive individuals.

Meanwhile the garment industry has pointed out the importance of quality control and has questioned whether the necessary consistency can be obtained in a so-called 'natural' product.

It is assumed that chemical plant process workers will operate the production facilities as they will replace plants traditionally operated by them. Assuming that operators will not be expected to walk more than 200 yards from the control room, the control rooms will have to be spaced 400 yards apart (that is, one per 33 acres). Over 750 control rooms will be required for a million pounds per year operation. Building costs will therefore be much higher than on a conventional plant, and may well make the new process uneconomic, especially now that the control rooms are being made stronger than in the past.

The greatest opposition to the new fibre is the result of the waste-disposal problems it will produce. Vast quantities of excreta will be deposited by the animals and will presumably have to be collected and dumped. Have the risks to health been fully considered? Will decomposition produce methane and a risk of explosions? What will happen when the animals become too old for productive use? It has been suggested that they might be used for human food and it is claimed that they are quite palatable after roasting. To quote the Director of the Centre for the Study of Strategic Perceptions, 'The suggestion is nauseating. Five thousand years after the dawn of civilization and 200 years after the industrial revolution, we are asked to eat the by-products of industrial production.'

# Pathological science

*General Electric Technical Information Series* No. 68-CO35, April 1968 (condensed). RBP from the General Electric Company.

At the General Electric Research Laboratory on December 18, 1953, Irving Langmuir, in a talk he called 'the science of things that aren't so', gave a colorful account of some pitfalls into which scientists may sometimes fall [1]. R N Hall has transcribed the essence of that talk from a poor-quality disk recording. Parts are summarized here.

## The Davis–Barnes effect, or counting hallucinations

In 1929, Bergen Davis from Columbia University gave a colloquium at the Research Laboratory, in Schenectady. He pictured a vacuum tube apparatus in which a beam of alpha particles from polonium could be counted by scintillations on a zinc sulfide screen, to which they were deflected by a magnetic field. Provision was made to have electrons of controlled speed move along with the alpha particles. The idea of the experiment was that if the electrons had the same speed as the alpha particles they might combine, giving singly-charged helium atoms, $He^+$. This was expected to occur for electrons which had been accelerated through 590 volts, and since any $He^+$ ions formed would be deflected only half as much as the remaining $He^{++}$ ions they would not reach the counting screen.

Davis and A H Barnes claimed they found such capture; indeed at 590 volts 80 per cent of the electrons combined with alpha particles. More remarkable: capture occurred for other voltages; those for which electrons in the beam had velocities which they would have in a Bohr orbit of $He^+$! The voltage peaks were sharp (claimed to be one hundredth of a volt wide, as read on a voltmeter with a 0–100 volt scale!), and for the right conditions the capture was always 80 per cent!

Whitney and Langmuir were skeptical of these results and also of several other aspects of Barnes's experiments. One difficulty arises because in the Bohr theory an electron coming from infinity has to give up half its energy to settle into the Bohr orbit. Since energy must be conserved, the electron has to radiate out an amount equal to the

297

energy it has left in the orbit. There was no evidence of radiation in the Barnes experiment. Further, when questioned about temperature dependence, Barnes said that even when the electron source (cathode) was at room temperature he observed recombination. Whitney objected that at such low current densities, about $10^{-20}$ amp, the electrons would be a long way apart. Barnes saw no difficulty: '...we now know that the electrons are waves. So the electron doesn't have to be there at all in order to combine with something. Only the waves have to be there and they can be of low intensity; the quantum theory causes all the electrons to pile in at just the right place where they are needed.'

Whitney and C H Hewlett, whose Geiger counters were offered to Barnes for more objective counting, visited the laboratory at Columbia University where Davis gladly demonstrated their apparatus, with the visitors participating in the measurements. This turned out to be a fiasco. But even Langmuir's 22-page letter to Davis pointing out that he was 'counting hallucinations' did not shake his confidence; he went on to read a paper on his work before the National Academy of Sciences.

A year and a half later, Davis and Barnes wrote a short article [2] in which they said they had not been able to reproduce the effect. 'The results reported in the earlier paper depended upon observations made by counting scintillations visually. The scintillations produced by alpha particles on a zinc sulfide screen are a threshold phenomenon. It is possible that the number of counts may be influenced by external suggestion or auto-suggestion to the observer...'

### Blondlot's rays

Not long after the discovery of X rays, R Blondlot, head of the physics department at the University of Nancy and member of the French Academy, announced that he had observed a new radiation having a variety of strange properties. He found that if you have a heated wire inside an iron tube fitted with an aluminium window, some rays come through that window (but not through the iron). In an almost dark room, what he called the N rays would make an object such as a sheet of paper much more visible. He found that the N rays could be stored in certain things, for example, a brick. Later he found that many kinds of things give off N rays. A human being gave off N rays, detectable to a dark-adapted observer, provided there was no *noise* in the darkened room!

Blondlot published numerous papers on these and other effects. One after another, other people did, too, most confirming Blondlot's results. At the urging of disbelieving friends, R W Wood, a renowned experimentalist from Johns Hopkins University, visited the laboratory where Blondlot and his assistant repeated some of their observations for him. One purported to refract the N rays through a large aluminium prism to show that they had several different components for which Blondlot was accurately measuring the refractive indices. After some time, Wood asked Blondlot to repeat some of the measurement, which he was glad to do. Unobserved, Wood had put the prism in his pocket. Yet Blondlot continued to recite measurements which checked perfectly with those made with the prism in place.

Wood's disclosure of his experience, even though he did not identify the laboratory by name, discredited Blondlot; there was also a rapid decrease in the number of 'confirmations' published by others. The Academy carried out its earlier plans to give Blondlot the Lalande prize of 20 000 francs and its gold medal but the presentation was made 'for his life's work, taken as a whole.' The tragic self-deception and exposure led to Blondlot's madness and death.

## Mitogenetic rays

Starting about 1923, hundreds of papers by Gurwitsch and others have been published on mitogenetic rays—rays given off by growing plants [3]. According to Gurwitsch, they would pass through quartz but not glass; they seemed to be some sort of ultraviolet light. Often studied with onion roots, the rays were claimed to be detectable with photographic film and with photocells and were claimed to influence the growth of neighboring roots. The effects were always borderline; less than half of the people who tried to repeat the reported experiments got confirmation of the effects.

## Allison effect

The Allison effect is one of the most interesting of all [4]. It makes use of the Faraday effect by which a beam of polarized light passing through a liquid has its plane of polarization rotated by a longitudinal magnetic field. Nicol prisms at each end of the test chamber permit measurement of the rotation. Allison devised a highly sensitive electrical way of measuring the time delay in the brightness of the transmitted light when a magnetic field was put on. He asserted that this delay was characteristic of the isotopes in solution. With it, a whole series of isotopes and new elements (Alabamine, Virginium...) were discovered by Allison.

When Wendell Latimer, head of the chemistry department at the University of California, expressed interest in the Allison effect, G N Lewis bet him ten dollars that 'you'll find there's nothing in it.' But after visiting Allison in Alabama, Latimer came back, set up the apparatus and made it work so well that Lewis paid him the bet. In fact, Latimer published a paper announcing his discovery of tritium $H^3$, using Allison's method.

Still there was enough skepticism about the Allison effect that the American Chemical Society decided it would not accept any more manuscripts on the subject. In 1932 R T Birge referred to the effect as 'Allison wonderland.' The paper by Jeppesen and Bell [5] was accepted by most as the coup de grace for the Allison effect. But some twenty years later, Allison, in retirement, and graduate students were still investigating the effect.

## Symptoms of pathological science

In false experiments where there is no dishonesty involved, people have been tricked into false results by a lack of understanding about how human beings can be led astray by subjective effects, wishful thinking, or threshold interactions. Cases of pathological science have sometimes attracted a great deal of attention, hundreds of papers have been published upon them; sometimes they last fifteen or twenty years before they gradually die away.

Langmuir suggested some warning symptoms of pathological science.

1 The maximum effect that is observed is produced by a causative agent of barely detectable intensity, and the magnitude of the effect is substantially independent of the intensity of the cause.
2 The effect is of a magnitude that remains close to the limit of detectability; or, many measurements are necessary because of the very low statistical significance of the results.
3 Claims of great accuracy.
4 Fantastic theories contrary to experience.
5 Criticisms are met by *ad hoc* excuses thought up on the spur of the moment.
6 Ratio of supporters to critics rises up to somewhere near 50% and then falls gradually to oblivion.

## References

[1] *Pathological Science* by I Langmuir, General Electric Research and Development Center, Technical Information Series No. 68-C-035 April 1968
[2] B Davis and A H Barnes 1932 *Physical Review* **37** 1368
[3] R W Wood 1904 *Nature* **70** 530
    R W Wood 1904 *Physikalische Zeitschrift* **5** 789
[4] F Allison and E S Murphy 1930 *Journal of the American Chemical Society* **52** 3796
    S S Cooper and T R Bell 1936 *Journal of Chemical Education* **13** 210, 278, 326
[5] W M Latimer and H A Young 1933 *Physical Review* **44** 690

# Atomic medicine

*John H Lawrence*

RBP from *California Monthly* December 1957 p 17 (excerpt).

It can be said that the new field of atomic medicine actually began at the University of California, where artificial radioactivity first became available for biological and medical research. Watching all the young men working around the cyclotron bombarding new targets and measuring the radiations with Geiger counters and Wilson cloud chambers, I was soon infected with the excitement of the early experiments. Very little was known of the biological effects of the neutron rays produced by the cyclotron, and this seemed an important place to start work.

For the neutron ray exposures in Berkeley we made a small metal cylinder to house a rat so that it could be placed close to the cyclotron. After placing the rat in position, we asked the crew to start the cyclotron and then turn it off again after the first two minutes. This two-minute exposure was arbitrary, since we had no basis for calculating how great a dose would produce an observable radiation effect on the animal. After the two minutes had passed, we crawled into the small space between the Dees of the 37-inch cyclotron, opened the cylinder, and found the rat was dead. Everyone crowded around to look at the rat, and a healthy respect for nuclear radiations was born. Now, of course, radiation protection measures are an integral part of all atomic energy research programs, but I think this incident of our first rat played a large part in the excellent safety record at the University. In fact, we have had no radiation cataracts among the early cyclotron workers. We discovered later that the rat's death had resulted from asphyxiation rather than radiation. But since our failure to aerate the rat chamber adequately had brought about such a salutory effect on the crew, the post-mortem report was not widely circulated.

The physicists were so busy and excited about their work they did not like to allow us exposure time for the animal experiments and thought us nuisances. One day as I walked by the cyclotron, a pair of pliers thoughtlessly left in my pocket were torn free by the intense magnetic field and flew into the Dees of the vacuum chamber of the cyclotron, putting it out of operation for three days. We were even less popular after this incident.

While you are up, bring me a grant.

*—Anon*

# A mathematician's miscellany

From *A Mathematician's Miscellany* by John E Littlewood (London: Methuen) 1960.

I once challenged Hardy to find a misprint on a certain page of a joint paper: he failed. It was his own name: 'G H Hardy'.

\*\*\*\*\*\*

A minute I wrote (about 1917) for the Ballistics Office ended with the sentence 'Thus $\sigma$ should be made as small as possible'. This did not appear in the printed minute. But P J Grigg said, 'what is that?' A speck in a blank space at the end proved to be the tiniest $\sigma$ I have ever seen (the printers must have scoured London for it).

\*\*\*\*\*\*

From an excellent book on Astronomy. 'Many of the spirals [galaxies], but very few of the ellipsoidals, show bright lines due, no doubt, to the presence or absence of gaseous nebulae.'
[This rich complex of horrors repays analysis. Roughly it is an illegitimate combination of the correct 'spirals show bright lines due to the presence...' and the incorrect 'ellipsoidals don't show bright lines due to the absence...']

\*\*\*\*\*\*

The spoken word has its dangers. A famous lecture was unintelligible to most of its audience because 'Hárnoo', clearly an important character in the drama, failed to be identified in time as $h\nu$.

\*\*\*\*\*\*

303

*Schoolmaster:* 'Suppose $x$ is the number of sheep in the problem'.
*Pupil:* 'But, Sir, suppose $x$ is not the number of sheep'.
[I asked Professor Wittgenstein was this not a profound philosophical joke, and he said it was.]

\*\*\*\*\*\*

'The surprising thing about this paper is that a man who *could* write it—would.'

\*\*\*\*\*\*

Landau kept a printed form for dealing with proofs of Fermat's last theorem. [It asserts that for an integer $n$ greater than 2 the equation $x^n + y^n = z^n$ is impossible in integers $x$, $y$, and $z$ all different from 0.] 'One page___, lines___ to___, you will find there is a mistake.' (Finding the mistake fell to the Privat Dozent.)

\*\*\*\*\*\*

It was said of Jordan's writings that if he had 4 things on the same footing (as $a$, $b$, $c$, $d$) they would appear as $a$, $M_3'$, $\epsilon_2$, $\pi_{1,2}''$.

\*\*\*\*\*\*

Veblen was giving a course of 3 lectures on 'Geometry of Paths'. At the end of one lecture the paths had miraculously worked themselves into the form

$$\frac{x - a}{l} = \frac{y - b}{m} = \frac{z - c}{n} = \frac{t - d}{p}$$

He then broke off to make an announcement about what was to follow, ending with the words 'I am acting as my own John the Baptist'. With what meaning I do not now recall (certainly not mine), but I was able to seize the Heaven-sent opportunity of saying 'Having made your own paths straight'.

\*\*\*\*\*\*

*Coincidences and improbabilities*

Improbabilities are apt to be overestimated. It is true that I should have been surprised in the past to learn that Professor Hardy had joined the Oxford Group. But one could not say the adverse chance was $10^6:1$. Mathematics is a dangerous profession; an appreciable proportion of us go mad, and then this particular event would be quite likely.

\* \* \* \* \* \*

A popular newspaper noted during the 1947 cricket season that two batsmen had each scored 1111 runs for an average of 44.44. Since it compared this with the monkeys' typing of *Hamlet* (somewhat to the disadvantage of the latter) the event is worth debunking as an example of a common class. We have, of course, to estimate the probability of the event happening at some time during the season. Take the 30 leading batsmen and select a pair $A$, $B$ of them. At some moment $A$ will have played 25 complete innings. The chance against his score then being 1111 is say 700:1. The chance against $B$'s having at that moment played 25 innings is say 10:1, and the further chance that his score is 1111 is again 700:1. There are, however, about $30 \times 15$ pairs; the total adverse chance is $10 \times 700^2/(30 \times 15)$, or about $10^4$. A modest degree of surprise is legitimate.

\* \* \* \* \* \*

There must exist a collection of well-authenticated coincidences, and I regret that I am not better acquainted with them. Dorothy Sayers in *Unpopular Opinions*, cites the case of two negroes, each named Will West, confined simultaneously in Leavenworth Penetentiary, U.S.A. (in 1903), and with the same Bertillon measurements. (Is this really credible?)

\* \* \* \* \* \*

Eddington once told me that information about a new (newly visible, not necessarily unknown) comet was received by an Observatory in misprinted form; they looked at the place indicated (no doubt, sweeping a square degree or so), and saw a new comet. (Entertaining and striking as this is, the adverse chance can hardly be put at more than a few times $10^6$.)

# Adventures of a cross-disciplinarian

*Craig F Bohren*

[Talk given to retired professors, Pennsylvania State University, 26 August 1991.]

RBP from C F Bohren.

I once belonged to the tribe of physicists, but was cast out and had to dwell among savages. I was like a young Englishman of a century ago who, lacking opportunities at home, had to leave Britain to take up a post in some remote outpost of empire. He went filled with jingoistic notions about taking up the white man's burden and bringing civilization to the natives. But in time, he learned their language and history and customs, and as a consequence came to see the natives of his own country in a different—and unflattering—light.

I am a physicist who has wandered into other fields and returns to tell of what I found in them and about my new perspectives on physics.

I was inspired to prepare this talk by the agitated state I was in when asked to speak. I recently had finished reviewing a manuscript submitted to a physics journal. The subject of this manuscript was what its authors called the blacksmith's paradox. This alleged paradox is the burning sensation a blacksmith feels in the hand grasping an iron bar when he thrusts its red-hot end into water. The authors, one of whom is a professor in a prestigious physics department, failed to detect a temperature rise at the cold end of a heated iron bar quenched in water, thereby casting doubts on the existence of the blacksmith's paradox.

306

I emphatically recommended that this manuscript not be accepted—and it was not—because its authors 'unwittingly tackled a problem in the physiology of sensation but treated it solely by the methods of physics. They have not recognized that the response of humans to heating and cooling is quite different from that of thermocouples.' I went further in my criticism and said that the authors 'nowhere indicate that they experienced what they purport to investigate ... it seems that they did not grasp the iron bar with their own hands, thrust it into cold water, and experience a burning sensation. ... This failure to experience the blacksmith's paradox directly indicates a lack of curiosity and an uncritical attitude toward experimentation.'

The lapses of individuals do not prove generalizations. I want to explore by adducing examples the errors inherent in a field, ones that, consciously or not, have been incorporated into its very bricks and mortar and hence cannot be rectified without tearing it down and building anew.

A few months ago, I gave a lecture in a physics department at another university. Afterwards, a member of the audience, presumably a physicist, and old enough to be a professor, asked me why the moon appears so large on the horizon. Although my talk did not touch on this question, I have become accustomed to being asked it. Indeed, I almost expect it. I tried my best to explain that the moon is not objectively larger on the horizon. Its apparent large size is a creation of the mind, not of the atmosphere. But my interrogator was difficult to satisfy. He had his own theory, namely, that the moon is magnified by gravitational bending of light. This ridiculous explanation stunned me even though I am inured to physicists' confident assertions that the enlarged horizon moon is a consequence of atmospheric refraction, which amuses me since most of them know nothing about atmospheric refraction even though they are lightning fast to invoke it when the need arises.

Gravitational bending of moonlight by the earth is

beyond the limits of detection. And bending of light by any physical mechanism has nothing to do with why the moon is perceived to be larger on the horizon. This is illusory, and has been known to be so for at least 1000 years. If you want to know about the moon illusion, ask a psychologist who specializes in perception. Don't ask a physicist.

Crepuscular rays are dark and light streaks, in a partly cloudy sky, converging toward the sun. Although the adjective crepuscular implies that these rays are seen at twilight, they are not restricted to the waxing or waning hours of daylight. Their appearance poses a problem. The sun is a nearly parallel source of light, subtending about half a degree at the earth. Yet the angular distance between crepuscular rays is much greater.

One of my students, Richard Herzog, wrote a term paper on crepuscular rays. He had shown photographs of them to 20 students. He said in his paper that 'I wanted to find out what these people, who are all in engineering or science majors, thought was the cause of the diverging rays. So that I could not be accused of tricking them, I phrased the question, "What do you think makes the rays *seem* to spread out?" All of the people I surveyed had a background in physics, which turned out to be their downfall. The vast majority ... attributed the spreading to some sort of bending of the light rays. Many responded that when the light passes through the gaps in the clouds, it is diffracted or dispersed like [the] light [that] passes through a pinhole. They were adamant that diffraction had to cause the spreading out of the rays.'

One student in 20 correctly recognized that the rays change in angular but not in linear separation. Crepuscular rays are an example of linear perspective, and have no more to do with diffraction or refraction than do converging railroad tracks.

If you want an explanation of crepuscular rays, don't ask a scientist, especially a physicist. You might do better asking an artist.

A common experience of a summer morning is to walk

across a lawn apparently wet with dew. But a closer look may reveal that all the drops are suspended from the tips of the blades. Alert students found an explanation of this in a book by a meteorologist whose critical faculties had been dulled by an overdose of physics:

'Dew forms primarily at the tips ... of blades of grass, because diffusion of water vapor from the air towards the tip can come from the largest range of angles. Also, the tips can radiate over a large range of angles towards the sky, without experiencing as much return radiation from other objects.'

This is an explanation that would satisfy physicists, but not botanists, because it is hopelessly wrong. The water drops are not dew drops, they are guttation. Dew is water condensed from the atmosphere whereas guttation originates in the soil. Guttation is most likely to occur on plants growing in warm, moist soil. At night, water continues to be forced up the stems of the plants by root pressure, but the stomates have closed so water is extruded from the end of the vein, the tip of the blade.

If you want to know about guttation, ask a botanist, not a physicist.

These are not isolated examples. I could give dozens more. And so, I daresay, could many of you. Every month I add to my stock of the wildly incorrect pronouncements of physicists, stated with the utmost confidence, even arrogance, on subjects about which they know nothing.

The moral of these few stories is not that physicists make errors. Making errors is healthy. To think is to err. But some errors are institutionalized. They have become inevitable, and such errors are not healthy.

The examples I have given embody at least four errors I consider to be part of physics in the sense that it is almost inevitable that physicists will make them.

The first error is implicit in a comment made by the late Helmut Landsberg in a review of a book on climate written by a physicist: 'The author is a physicist and has the attitude, which has prevailed in that profession for some time, that education in physics makes you an expert on almost anything.'

The second fundamental flaw or error in the physics edifice is that physicists who tackle problems outside their competence feel under no obligation to read any papers from the fields in which these problems lie or to seek advice from scientists in these fields.

The third error is the notion that the objective world is all that matters, the subjective feelings of human beings either should be denied or ignored.

And finally, there is the notion that there is no need to experience directly whatever you are investigating even if you can.

Why have I gone to the trouble to do so much physics bashing? Rightly or wrongly, physics is looked upon as the model to which all other sciences must aspire. Lord Rutherford is quoted as saying 'Science is *physics*; everything else is stamp collecting'. In many fields we see the spectacle of scholars scrambling to ape the methods of physicists, often with ludicrous results, as for example, in sociology.

One of the most widely quoted statements about the superiority of quantity to quality—of physics to stamp collecting—was made by Lord Kelvin a century ago: 'When you can measure what you are speaking about, and express it in numbers, you know something about it; but when you cannot measure it, when you cannot express it in numbers, your knowledge is of a meager and unsatisfactory kind; it may be the beginning of knowledge, but you have scarcely, in your thoughts, advanced to the stage of science.'

I have seen and heard this half truth so many times that I finally decided to fling it back in the faces of those who parrot it. Kelvin spent part of his career battling geologists about the age of the earth. They had their estimates based on their methods, he had his estimates based on his calculation of the cooling of the earth. Although Kelvin expressed his results 'in numbers', they were wrong numbers. He was wrong because he had not accounted for radioactivity, which had not yet been discovered. What disturbs me is not that he was wrong but that he wouldn't concede to geologists that they might be right.

The best refutation of Kelvin is to note that the answers to the most important questions are *not* quantitative. I'll give one example: Is there life of any kind beyond our planet? Yes or no is all that is wanted. If the answer turns out to be yes, it will have profounder consequences that any quantitative knowledge ever discovered.

The prestige of physics among the non-scientific public is to a large extent a consequence of the bombs dropped on Hiroshima and Nagasaki. Physicists don't like to admit this but it is nevertheless true. Another reason for the prestige of physics is that it has been a source of gadgets such as transistors and lasers.

Among intellectuals the prestige of physics originates partly from its success in achieving its aims. This success, however, has been obtained by applying extremely complicated methods to extremely simple systems, a distinction rarely noted. The electrons in copper may describe complicated trajectories but this complexity pales in comparison with that of an earthworm. Physics uses complicated mathematics, which can be used to make people feel inadequate, inferior, and therefore to intimidate them.

Another reason for the prestige of physics among intellectuals is the faith we have in reductionism. The most self-serving reductionist statement I have found was made in 1966 by Wolfgang Panofsky, then director of the Stanford Linear Accelerator: 'All other physical, and probably all life sciences must ultimately rest on the findings of elementary particle physics. It would indeed violate all our past experiences in the progress of science if nature had created a family of phenomena which governs the behavior of elementary particles without at the same time establishing any links between these phenomena and the large-scale world which is built from those very particles.'

In fact, 'all our past experiences' point to just the opposite conclusion, as noted by John Waymouth, who divided physics into three basic areas: electron-volt—or eV—physics, on the atomic and molecular scale; MeV – GeV physics, on the scale of nuclear and sub-nuclear particles; GeV – TeV physics, or high-energy physics, on the scale of subnuclear matter.

eV physics has given us electric light and power, radio, television, lasers, and much more. It is the 'core science of the modern world'.

MeV – GeV physics has given us radioisotope analysis, part of medical physics, and nuclear power.

'High-energy physics has to date given us nothing.'

'eV physics is the science of things that happen on Earth; MeV – GeV physics is the science of things that happen in the Sun; TeV physics has not happened anywhere in the universe since the first few milliseconds of the Big Bang.'

Panofsky's self-serving statement is a variation on the following faulty sequence: physics is really just quantum mechanics; chemistry is really just physics; biology is really just chemistry; medicine is really just biology. Therefore, in principle finding a cure for cancer is really just a problem in quantum mechanics.

In the time remaining to me, let me share with you some of my thoughts on the reductionist – fundamentalist view of science. The people who are most actively trying to determine the fundamental laws of nature are at the same time the least interested in deducing the consequences of those laws and the most contemptuous of those who are. I used to think that physics is the study of the physical world, including that which we immediately experience. Yet every year I become more dismayed by the attitude of physicists, especially those of the nuclear and high-energy stripe, who not only know little about the world around them but often are openly contemptuous of it. It sometimes seems that physics is a monastery into which those who hate the world escape.

Those who are hot on the trail of the ultimate theory of everything are proceeding on the expectation that matter has an ultimate structure, that there is an end to elemen-

312

tary particles. Yet there is not a scrap of evidence to support this. It is just as likely that matter can be subdivided indefinitely, that protons are made of something, which is made of something else, which is made of something else, and so on *ad infinitum*. But experience does tell us that it takes more energy, hence more money, to peel off each successive layer of the atom. There are practical limits to how deep we can dig. And there are also philosophical limits: if matter really is like an onion with an infinite number of layers, it is pointless to proceed.

It recently occurred to me that there are some parallels between the new physics and ideology as described by Hannah Arendt in *The Origins of Totalitarianism*. According to her, ideologies are 'isms which to the satisfaction of their adherents can explain everything and every occurrence by deducing it from a single premise'. An 'ideology ... claims to possess either the key to history, or the solution for all the "riddles of the universe", or the intimate knowledge of the hidden universal laws which are supposed to rule nature and man.' We usually think of ideologies as belonging to the political or spiritual realms. Yet Arendt's description of an ideology exactly fits physics as practiced by those who believe that they are on the threshold of the theory of everything.

The physics church today is dominated vocally, although not numerically, by what I call the fundamentalists, those who believe that the only proper task for physicists is the search for fundamental laws. Yet most physicists are not engaged in this search. They, like most other scientists, seek to understand patterns in nature, ones they find intellectually or aesthetically pleasing, or even useful in some sense. Moreover, the fundamentalists themselves have promised us that with the expenditure of only a few more billions of dollars, all the fundamental laws will have been discovered. If we accept their view of physics, we must conclude that most physicists are not doing physics and, even if they are now, they won't be for long. When all the fundamental laws of physics are finally wrung from nature, there will be one last trip to

Stockholm and then a retreat to studies for the writing of memoirs.

An alternative definition of physics is that it is a search for and unraveling of patterns in nature and the links among them with the aid of what has come to be called physical laws. With this interpretation, physicists will always be faced with challenges, for although there may be an end to fundamental laws, there seems to be no end to patterns in nature. When fundamentalism dies, ecumenism will rise to take its place.

If we accept the view that not only physics but all science is essentially a search for patterns, then all scientists are linked leaving the fundamentalists in the isolation they so richly deserve.

---

# Mathematical motivation through matrimony

*Martin D Stern*

From *Mathematics Magazine* **63.4** October 1990 pp 231–3

**Introduction.** At the end of the Book of Common Prayer the following table of prohibited degrees of marriage is given.

Though the prohibited marriages are arranged in parallel columns for men and women, it is not immediately obvious that there are no other implicit prohibitions. For example, according to line 15 in the first column, a man may not marry his mother's father's wife (obviously after her husband's demise!), and the question that might cross the reader's mind is whether this woman is explicitly prohibited from marrying him. In order to determine whether this is so, the relationship must be inverted and the purpose of this article is to describe a mathematical notation based on binary numbers that facilitates this inversion process.

A TABLE OF
KINDRED AND AFFINITY
Wherein whosoever are related are forbidden by the Church of
England to marry together

*A man may not marry his:*

1 Mother
2 Daughter
3 Father's mother

4 Mother's mother
5 Son's daughter
6 Daughter's daughter

7 Sister
8 Father's daughter
9 Mother's daughter

10 Wife's mother
11 Wife's daughter
12 Father's wife

13 Son's wife
14 Father's father's wife
15 Mother's father's wife

16 Wife's father's mother
17 Wife's mother's mother
18 Wife's son's daughter

19 Wife's daughter's daughter
20 Son's son's wife
21 Daughter's son's wife

22 Father's sister
23 Mother's sister
24 Brother's daughter
25 Sister's daughter

*A woman may not marry her:*

1 Father
2 Son
3 Father's father

4 Mother's father
5 Son's son
6 Daughter's son

7 Brother
8 Father's son
9 Mother's son

10 Husband's father
11 Husband's son
12 Mother's husband

13 Daughter's husband
14 Father's mother's husband
15 Mother's mother's husband

16 Husband's father's father
17 Husband's mother's father
18 Husband's son's son

19 Husband's daughter's son
20 Son's daughter's husband
21 Daughter's daughter's husband

22 Father's brother
23 Mother's brother
24 Brother's son
25 Sister's son

The notation was originally developed for analysing problems in Jewish marriage law [1], where the prohibited degrees are only stated from a man's standpoint in the standard codes. The problem considered was to list the corresponding prohibitions for a woman. In the latter system there was not a symmetrical relationship between the sexes. For example a man was allowed to marry his niece but not his aunt so, consequently, a woman was prohibited from marrying her nephew but not her uncle.

In the problem here considered we shall show that there are no further prohibited marriages other than those explicitly stated and hence that there is an element of symmetry between the sexes regarding the restriction in the choice of a marriage partner.

**Notation.** We shall use the single digits 1 and 0 to denote male and female, respectively, and define the two digit codes for relationships between individuals:

| | |
|---|---|
| 00 | spouse |
| 01 | parent |
| 10 | child |
| 11 | sibling |

The choice of a particular code is, of course, arbitrary except that a symmetric relationship such as spouse or sibling must consist of two identical digits whereas parent or child may not. The reason for this will become evident when the problem of inverting relationships is discussed below.

To establish the relationship between one person and another, we write a digit for that person followed by a sequence of three digits, the first two representing the relationship and the third the sex of the second person, e.g., a man's wife would be expressed as 1 000.

Some sequences are inadmissible since, for example, only heterosexual marriages are officially sanctioned by the Church of England, i.e., 1001 would be inadmissible.

More complicated relationships can be expressed by appending further triplets. So, for example, a man's wife's father's brother can be written as

$$1\ 000\ 011\ 111.$$

Conversely any sequence of $3n + 1$ binary digits can be interpreted as a relationship between two people. For example 1011000101 can be analysed as

| 1 | 011 | 000 | 101 |
|---|---|---|---|
| a man's | father's | wife's | son |

i.e., a step brother. We shall assume that relationships will always be expressed in the most economical manner so that this is distinct from a brother 1 111 or a (paternal) half brother, i.e., a man's father's son which would be 1 011 101.

**Inversion of relationships.** Using this notation it is easy to find the inverse of a given relationship by merely reversing the order of the digits and regrouping them as a single digit followed by a sequence of triplets. For example, a man's mother's father's sister's husband's mother would be expressed as

1 010 011 110 001 010.

When this procedure is followed we obtain

0 101 000 111 100 101

which can be readily interpreted as a woman's son's wife's brother's daughter's son.

When this notation is applied to the table of kindred and affinity we obtain the results in the table below. In it we have given the prohibitions as listed for a man followed by its code, the inverse code and the corresponding prohibition for a woman. Since every prohibited relationship for a woman appears in this final column, we see that there are no further prohibitions implicit in the table. This contrasts with other relationship problems to which this notation has been applied, such as the case of the prohibition in rabbinic law of consanguinous relatives from acting as witnesses [2].

**Conclusion.** While the notation presented here is not claimed to be particularly profound it might provide a useful tool for motivating pupils to take a greater interest in mathematics by emphasizing the non-numerical aspects of the subject. In England, where we have in our inner cities a considerable mix of people of different religious and ethnic backgrounds, this notation has been used as a basis for projects in which the students have analysed one another's traditions and, hopefully, come to a closer understanding both of their fellows and the value of mathematics.

317

| Male | Code | Inverse | Female |
|------|------|---------|--------|
| *A man may not marry his:* | | | *A woman may not marry her:* |
| 1 Mother | 1 010 | 0 101 | 2 Son |
| 2 Daughter | 1 100 | 0 011 | 1 Father |
| 3 Father's mother | 1 011 010 | 0 101 101 | 5 Son's son |
| 4 Mother's mother | 1 010 010 | 0 100 101 | 6 Daughter's son |
| 5 Son's daughter | 1 101 100 | 0 011 011 | 3 Father's father |
| 6 Daughter's daughter | 1 100 100 | 0 010 011 | 4 Mother's father |
| 7 Sister | 1 110 | 0 111 | 7 Brother |
| 8 Father's daughter | 1 011 100 | 0 011 101 | 8 Father's son |
| 9 Mother's daughter | 1 010 100 | 0 010 101 | 9 Mother's son |
| 10 Wife's mother | 1 000 010 | 0 100 001 | 13 Daughter's husband |
| 11 Wife's daughter | 1 000 100 | 0 010 001 | 12 Mother's husband |
| 12 Father's wife | 1 011 000 | 0 001 101 | 11 Husband's son |
| 13 Son's wife | 1 101 000 | 0 001 011 | 10 Husband's father |
| 14 Father's father's wife | 1 011 011 000 | 0 001 101 101 | 18 Husband's son's son |
| 15 Mother's father's wife | 1 010 011 000 | 0 001 100 101 | 19 Husband's daughter's son |
| 16 Wife's father's mother | 1 000 011 010 | 0 101 100 001 | 20 Son's daughter's husband |
| 17 Wife's mother's mother | 1 000 010 010 | 0 100 100 001 | 21 Daughter's daughter's husband |
| 18 Wife's son's daughter | 1 000 101 100 | 0 011 010 001 | 14 Father's mother's husband |
| 19 Wife's daughter's daughter | 1 000 100 100 | 0 010 010 001 | 15 Mother's mother's husband |
| 20 Son's son's wife | 1 101 101 000 | 0 001 011 011 | 16 Husband's father's father |
| 21 Daughter's son's wife | 1 100 101 000 | 0 001 010 011 | 17 Husband's mother's father |
| 22 Father's sister | 1 011 110 | 0 111 101 | 24 Brother's son |
| 23 Mother's sister | 1 010 110 | 0 110 101 | 25 Sister's son |
| 24 Brother's daughter | 1 111 100 | 0 011 111 | 22 Father's brother |
| 25 Sister's daughter | 1 110 100 | 0 010 111 | 23 Mother's brother |

## References

[1] J Cashdan and M D Stern 1987 Forbidden marriages from a woman's angle *Math. Gaz.* **71** No. 456
[2] M D Stern 1987 Consanguinity of witnesses—a mathematical analysis *Teaching Mathematics and Its Applications* Vol 6.2

# This is the Theory Jack Built

*This is the Theory Jack built*

*This is the Flaw*
*That lay in the Theory Jack built.*

*This is the Mummery*
*Hiding the Flaw*
*That lay in the Theory Jack built.*

*This is the Summary*
*Based on the Mummery*
*Hiding the Flaw*
*That lay in the Theory Jack built.*

*This is the Constant K*
*That saved the Summary*
*Based on the Mummery*
*Hiding the Flaw*
*That lay in the Theory Jack built.*

*This is the Erudite Verbal Haze*
*Cloaking Constant K*
*That saved the Summary*
*Based on the Mummery*
*Hiding the Flaw*
*That lay in the Theory Jack built.*

*This is the Turn of a Plausible Phrase*
*That thickened the Erudite Verbal Haze*
*Cloaking Constant K*
*That saved the Summary*
*Based on the Mummery*
*Hiding the Flaw*
*That lay in the Theory Jack built.*

# Warnings

*Gloom of Thermodynamics—*

You cannot win, you can only break even.
You can only break even at the absolute zero.
You cannot reach absolute zero.
Conclusion: You can neither win nor break even.

*Anon*


Whenever you look at a piece of work and you think
the fellow was crazy, then you want to pay some
attention to that. One of you is likely to be, and you
had better find out which one it is. It makes an awful lot
of difference.

*Charles F Kettering*

# Pickles will kill you!

*Anon*

Communicated by D H Janzen, *Journal of Insignificant Research* **6** (1) July 8, 1979 p 13. RBP of Blackwell Scientific Publications, Inc.

Every pickle you eat brings you nearer to death. Amazingly, the 'Thinking man' has failed to grasp the terrifying significance of the term 'in a pickle'. Although leading horticulturists have long known that *Cucumis sativus* possesses an indehiscent pepo, the pickle industry continues to expand.

Pickles are associated with all the major diseases of the body. Eating them breeds wars and Communism. They can be related to most airline tragedies. Auto accidents are caused by pickles. There exists a positive relationship between crime waves and consumption of this fruit of the cucurbit family.

For example:

- Nearly all sick people have eaten pickles. The effects are obviously cumulative.
- 99.9% of all people who die from cancer have eaten pickles.
- 100% of all soldiers have eaten pickles.
- 96.8% of all Communist sympathizers have eaten pickles.
- 99.7% of the people involved in air and auto accidents ate pickles within 14 days preceding the accident.
- 93.1% of juvenile delinquents come from homes where pickles are served frequently

Evidence points to the long-term effects of pickle-eating:

- Of the people born in 1859 who later dined on pickles, there has been a 100% mortality.
- All pickle eaters born between 1869 and 1879 have wrinkled skin, have lost most of their teeth, have brittle bones and failing eyesight—if the ills of eating pickles have not already caused their death.
- Even more convincing is the report of a noted team of medical specialists: rats force-fed with 20 pounds of pickles per day for 30 days developed bulging abdomens. Their appetites for WHOLESOME FOOD were destroyed.

322

In spite of all the evidence, pickle grower and packers continue to spread their evil. More than 120,000 acres of fertile U.S. soil are devoted to growing pickles. Our *per capita* consumption is nearly four pounds.

Eat orchid-petal soup. Practically no one has as many problems for eating orchid-petal soup as they do with eating pickles.

---

## Silly asses

Isaac Asimov in *Future* February 1958. RBP of the US Junior Chamber of Commerce.

Naron of the long-lived Rigellian race was the fourth of his line to keep the Galactic records.

He had the large book which contained the list of the numerous races throughout the Galaxies that had developed intelligence, and the much smaller book listing those races that had reached maturity and had qualified for the Galactic Federation. In the first book, a number of those listed were crossed out; those that, for one reason or another, had failed. Misfortune; biochemical or bio-physical shortcomings; social maladjustment, etc, took their toll. In the smaller book, however, no member listed had yet blanked out. And now Naron, large and incredibly ancient, looked up as a messenger approached.

'Naron', said the messenger. 'Great One!'

'Well, well, what is it? Less ceremony.'

'Another group of organisms has attained maturity.'

'Excellent. Excellent. They are coming up quickly now. Scarcely a year passes without a new one. And who are these?'

The messenger gave the code number of the Galaxy and the co-ordinates of the world within it. 'Ah, yes', said Naron. 'I know the world'. And in flowing script, he noted it in the first book and transferred its name into the second—using, as was customary, the name by which the planet was known to the largest fraction of its population. He wrote:

Earth.

He said, 'These new creatures have set a record. No other group has passed from intelligence to maturity so quickly. No mistake, I hope.'

'None, sir', said the messenger.

'They have attained thermonuclear power, have they?'

'Yes, sir.'

'Well, that's the criterion.' Naron chuckled. 'And soon their ships will probe out and contact the Federation.'

'Actually, Great One', said the messenger, reluctantly, 'the Observers tell us they have not yet penetrated space.'

Naron was astonished. 'Not at all? Not even a space station?'

'Not yet, sir.'

'But if they have thermonuclear power, where then do they conduct their tests and detonations?'

'On their own planet, sir.'

Naron rose to his full twenty feet of height and thundered, 'On their **own** planet?'

'Yes, sir.'

Slowly, Naron drew out his stylus and passed a line through the latest addition in the smaller book. It was an unprecedented act, but, then, Naron was very wise and could see the inevitable as well as anyone in the Galaxy.

'Silly asses', he muttered.

# Fermi on the importance of being quantitative

*Albert A Bartlett*

RBP from *Am. J. Phys.* **44** (3) 235 (1976).

At one point in a seminar on nuclear physics at Los Alamos in 1943, Enrico Fermi was discussing the way in which a cross section increased rapidly with energy and then leveled out. To illustrate this, he drew Figure 1 on the blackboard, emphasizing that one could estimate the numerical value of the cross section on the plateau of the curve.

Figure 1.

He paused. Then he stepped back from the board, took a small slide rule from his pocket, and made a quick calculation. This led him to say 'It's not that high', and he erased the upper part of the curve and redrew it as shown in Figure 2.

Figure 2.

For a moment the room was filled with amused silence followed by an avalanche of laughter. Fermi looked puzzled at first; then, smiling briefly in recognition, he continued the lecture.

# Dread of nakedness

RBP from Adrian Berry *High Skies and Yellow Rain,* © The Daily Telegraph plc, 1983.

*Approach thou like the rugged Russian bear,*
*The arm'd rhinoceros, or the Hyrcan tiger;*
*Take any shape but that.*

*— Macbeth*

What news of black holes? It has been some years since these objects burst upon public consciousness. Most people have now accepted the fantastic-seeming prediction of giant stars that have ceased to shine because their gravitational fields are so strong that no light can escape from them.

But black holes may yet have a surprise for us that is to be dreaded rather than hoped for. A situation could arise in which the discovery of one of these objects posed an actual threat to civilisation—not of a physical character, but rather from the information that it revealed.

Consider the two basic components of a black hole. Take first the 'singularity' within it, where the volume of the crushed star has been reduced to zero while its original mass has barely changed. It is a place of gigantic density in which not even subatomic particles can exist.

The second is the 'event horizon', that hides the singularity from outside eyes. Every object in space has a certain speed, an 'escape velocity', that must be attained if anything is to escape from it. From Earth, for example, a rocket must rise at 25 000 mph to defeat the pull of gravity and get into orbit. The greater a planet's mass, the higher its escape velocity. Now a black hole might have a mass three million times greater than the Earth's. Its escape velocity would exceed the speed of light, 670 million mph. Nothing, not even light itself, can normally escape from a black hole.

And so, as a black hole forms, an event horizon forms around it, and the light of its singularity, although still burning within, is to a distant observer snuffed out, like a car headlight extinguished in the night. All information about the singularity is hidden, only to be explored indirectly, by mathematical speculation.

326

But scientists see as a threat to their profession—yes, a threat—the possibility that in sufficiently extraordinary conditions, involving swirling deformities of mass, a singularity might form *without an event horizon*. The singularity would be 'naked'. We would be able to see into it. We could observe for ourselves the total breakdown of the laws of physics. We would see what Professor Paul Davies calls the 'edge of infinity'.†

Why would this situation threaten anyone? It is because *there would be no way whatever of predicting what would emerge from the singularity*. It would be the gateway into this universe from another, from a place, perhaps, where the laws of space, time and matter were different from our own.

The discovery of a naked singularity, says Professor Davies, would be a 'desperate crisis'.

'One can imagine', he explains, 'an Alice-in-Wonderland world in which the singularity coughs out all manner of weird and wonderful preformed objects, stars, planets, people, computers, copies of encyclopaedias. In a lawless universe (which this would be) anything goes.'

But how could this be? Surely, it will be objected, the singularity could only spew forth the relics of what had formed it. But this is not the case. The professor and his colleagues are explicit on the point. Up to the time when the singularity is formed, the behaviour of the doomed star is determined by prevailing physical conditions. But once the singularity takes shape, the influence of these conditions ceases. The thing becomes wholly alien.

If physical laws are seen to break down, the reputation of physicists breaks with them. A world where events could not be predicted would be no place for rationality or logic. The hallowed principle of cause and effect would be destroyed. The scientist would have to step down from his lofty seat and surrender his place to the necromancer.

---

†Paul Davies, Professor of Theoretical Physics at Newcastle University, describes naked singularities and their consequences in his book *The Edge of Infinity: Beyond the Black Hole* (Oxford University Press).

The last known naked singularity burst forth about 15 000 million years ago, and it did indeed emit marvellous objects, which later metamorphosed into typewriters, chessboards and princesses. I refer, of course to the Big Bang, the explosion from nowhere which began the universe. It was not only matter, space and time which started with the Big Bang. So also did order and logic; before it, none of these five things existed.

What happened once could happen again. Even on a small scale, a recurrence of the Big Bang could bring less welcome phenomena. There are many distant, violently-radiating objects in space that might turn out to be naked singularities. Let us hope none does.

## Unnatural 'nature'

RBP from Adrian Berry *High Skies and Yellow Rain*, © The Daily Telegraph plc, 1983.

Why do so many people decline to take any interest in science or engineering? The usual answer, that they find these subjects hard to understand because they tend to be written about in difficult jargon, may be misleading. The real objection, it seems to me, is not so much to the jargon but to the vile English that accompanies it.

It is hard enough having to wade through a text that contains such unfamiliar words as 'maser', 'pulsar' and 'metagalaxy'. But it becomes ten times harder when such words and phrases are abbreviated into groups of initials and dots; when they are strung into sentences of intolerable length, not even broken by colons or semi-colons, and when the main verb has either been omitted or has been cunningly hidden in a subordinate clause.

Where, then, should we seek examples of bad scientific writing? Perhaps most of all in *Nature*, the British journal that was founded 114 years ago with the stated purpose of making science intelligible to ordinary people. Today, it does the opposite. Not only do most of the articles in *Nature* seem to non-scientific people as if they were written in a foreign language, but they are often incomprehensible to scientists in the field under discussion.

Since *Nature*, although privately owned, is widely regarded as the 'official' organ of British science, I have taken the liberty of picking from recent issues sentences which might have given Fowler apoplexy. The italics are my own:

> *It has become generally accepted that superluminal motion is observed in the nuclei of some radio sources.*

Why not simply say: 'Certain objects in the centres of exploding galaxies appear to travel faster than light'? It would attract much wider interest.

> *An inspirational 1972 paper by Niles Eldredge and Stephen Gould has provoked fruitful empirical work and active debate about whether evolution proceeds mostly by phyletic gradualism or by punctured equilibrium (Stanley's 'rectangular' evolution).*

My translation: 'People are wondering whether evolution proceeds gradually or by sudden jumps'. The adjectives 'inspirational', 'fruitful' and 'empirical' are clearly redundant. And this is an example of the confusion, typical among some scientists, of trying to cram too much information into one sentence. The references to Eldredge and Gould should have been postponed until later. And there should have been a sentence of its own to explain what Stanley means by calling evolution 'rectangular'.

> *It is a direct consequence of the Einstein equivalence principle (EEP) that all atomic clocks will run at the same rate if situated at the same point in space – time.*

We hardly need Einstein to tell us that two accurate clocks in the same room will run at the same speed.

> *The Arabian Peninsula often appears totally aseismic on maps of world seismicity, reflecting the apparent absence of earthquakes during the 20th century and particularly since the improvement of the worldwide seismological network after the 1960s.*

Why not just say: 'The Arabian Peninsula appears to have been free of earthquakes'?

*For at least the past 10 years, a mild controversy has sim-mered in meteoritical and astrophysical circles regarding the strange distribution of xenon isotopes found in certain meteorites.*

Having rewritten this to say: 'People have been wonder-ing for 10 years in an idle sort of way why some meteorites contain xenon', one must severely criticise it. Who else but those in 'meteoritical and astrophysical circles' would be arguing about meteorites. Linguists? Sanitary engineers? Surely not. It is unnecessary verbiage. And the statement that the controversy is 'mild' is sufficient to kill any interest in the subsequent discussion.

*Measurements of stratospherical minor constituent gas con-centrations are important for the elucidation of the complex chemistry amid the evaluation of the predictions presented by theoretical models regarding the short and long term future of this important atmospheric region.*

I would have said: 'To understand the stratosphere one should study its gases'.

*The reappearance of Van Nostrand's* Scientific Encyclopedia *in a new edition (the sixth) only seven years after the last edition, combined with its sheer bulk (3,067 pages, 700 more than the fifth edition), provides welcome assurance that this very considerable work is now a perma-nent feature of the scientific landscape ... the Mont Blanc, if not the Everest (the 15-volume McGraw Hill* Encyclopedia of Science and Technology *is surely that), of scientific encyclopedias.*

I think this run-amok sentence means: 'Van Nostrand's *Scientific Encyclopedia* is nearly as good as McGraw Hill's, but it's hard to be absolutely sure.

Some people would argue, in perfect seriousness, that contributors to *Nature* must write in this convoluted way to preserve its editorial character. Even if very few under-stand it, they say, it sounds magisterially impressive, and that is what matters.

The argument does not seem very sensible. *Nature* has a weekly circulation of nearly 26 000, while comparable American scientific journals have circulations of nearly a million—because they are so much better written. As a result, most Americans seem to know much more about science than most Britons.

---

# An economist treats mathematics poetically

*Kenneth E Boulding*

RBP from X Cantos, *Michigan Quarterly Review* VIII, 1, Winter 1969.

## Canto I

*How pleasant it can be to sit*
*And contemplate the infinite!*

## Canto II

*The integers march by in fine*
*Unending, but still counted, line*
*And in between them march the class*
*Of fractions in a solid mass.*
*Then in the holes that don't exist,*
*Between the fractions, we must list*
*Uncountable irrational hosts*
*Of infinitely slender ghosts,*
*While somewhere in the endless sky*
*Imaginary Numbers fly*
*Illuminated by the ethereal sun*
*Of the square root of minus one.*

## Canto XII

*A matrix is a set of roles*
*Arranged like banks of pigeon-holes.*
*It is a box of empty boxes,*
*Filled with nothing but paradoxes.*
*Their mathematical operations*

*Involve elaborate permutations*
*In which the x's and the i's and j's*
*Dance algebraic ballets,*
*And in the course of their gyrations*
*Solve simultaneous equations.*

### Canto XIII

*It gives a piece a certain unction*
*When each relation's called a Function,*
*And if prestige is what we seek,*
*We write the functions out in Greek.*
*These intellectual acrobatics*
*Are fun, but are not mathematics,*
*For there are subtle traps immense*
*Where all that's symbol is not sense.*

---

## There was a young chemist ...

Contributed by the author, Frank Hawke.

*There was a young chemist in Ealing*
*Who with trinotrophenol was dealing.*
*But, he added red lead*
*And, the truth must be said,*
*They found him a splash on the ceiling.*

---

## Put your brains in your pocket

RBP of Arthur W Hoppe *The San Francisco Chronicle.* © 1974 Chronicle Publishing Co.

At last the American ideal of true equality for one and all is in sight. The harbinger is that latest rage, the pocket calculator.

In Berkeley, for example, the Board of Education has approved buying pocket calculators for tots who have difficulty learning to multiply. This way, they won't have to learn to multiply. Yet they'll be able to go forth and multiply perfectly for the rest of their lives—or at least until their batteries go dead.

The next step is obvious: a pocket computer with a miniaturized memory bank capable of storing billions of facts and the ability not only to multiply but to analyse, deduce, and program solutions to every conceivable problem.

Actually, just such a device was developed as long ago as 1938 by the famed electronics wizard, Dr Wolfgang von Houlihan. Realizing the tremendous potential for human equality inherent in his invention, Dr von Houlihan decided to test it out first on his only son, Egbert.

Egbert was an ideal subject. It was not that he lacked the intelligence to do well in his school down the block. It was that he lacked the intelligence to find his school down the block.

But after weeks of patient instruction, his father was able to teach Egbert which buttons to push when. The change in him was startling.

With billions of facts at his fingertip, he naturally quit school. And, knowing everything, he naturally read nothing. And yet, unschooled and unread, he whizzed through life.

His employers were amazed by his incredible knowledge, his cool deductions, his brilliant analyses and his invariably perfect solutions. At thirty-five, he became head of General Conglomerated, Inc.

'I got my brains from my father', he would say modestly when complimented. And then he would hum a few bars of his favorite song, 'I've Got a Pocketful of Brains'.

His wit and erudition made him a hit at cocktail parties. He always said the right thing, did the right thing, voted for the right candidate and never once forgot his mother's birthday.

He was the perfect businessman, the perfect companion, the perfect citizen and—after he had computed the proper steps to sweep the beautiful Millicent Oleander off her feet—the perfect husband. While Millicent found Egbert singularly uncommunicative in the shower or in bed, she was perfectly happy with the perfect spouse who never once forgot their anniversary.

Needless to say, Egbert's father was overjoyed with the success of the experiment. 'Just think', he cried. 'When all people carry their brains in their pockets, all will be not only equal but perfect!'

Dr von Houlihan was about to unveil his device for the perfection of mankind, when The Catastrophe struck. Afterward, he destroyed all his blueprints, muttering, 'Equality's nice, but maybe we ought to just struggle along with what we've got.'

What happened, of course, was that one morning while her husband was in the shower, Millicent sent his pants to the cleaners. And Egbert lost his mind.

---

## Guests and hosts

### *You can dress a biologist up, but you can't take him out.*

RBP from Milton Love *Natural History* **88** 84, June/July 1979.

I have reluctantly come to the conclusion, after much soul-searching and not a little pain, that biologists make poor dinner guests. Not that we eat with our hands, fail to bathe, or lack other social graces. The problem is that our conversation tends to encompass topics not usually included in dinner-table repartee. Just as lawyers may animatedly discuss a particularly interesting case, biologists can wax poetic about cockroaches, cannibalism, and the like.

My wife has become more or less resigned to such behavior. She was introduced to it when, in a fit of romanticism, I named a parasite after her. She had her choice of organisms: one parasitized the gall bladders of certain fishes; the other, their urinary bladders. (She picked the urinary bladder parasite—*Davisia reginae*.)

Even she, however, eventually drew the line. During dinner one night, after discoursing on *kura* (a viral disease found among New Guineans, caused by ritual brain cannibalism) and the pearlfish (which lives in the anus of sea cucumbers), I happened to mention *Oikopleura*, a small planktonic animal that lives in mucus 'houses' which also serve as filters of food material.

'That's it!' she declared. 'There will be no discussion of mucus at the dinner table'. Even thoroughly inured non-biologists have their limits.

The real problem arises when we are invited to gatherings. This was made abundantly clear during a Passover Seder my wife and I recently attended. A seder is a ceremonial dinner celebrating the escape of Moses and his flock from Egypt. I often find the conversations at these affairs to be dull, and consequently, I spend much of these evenings in a sort of stupor.

On this occasion, the conversation swung around to the kosher dietary laws. Perking up, I pointed out that the biblical prohibition against eating pork was probably not based on the prevention of trichinosis (as is commonly believed), for trichinosis, a parasitic disease contracted by eating nematode-infected pork, is only rarely found in the Middle East.

'Trichinosis', I reflected, warming to my subject, 'is an interesting disease. Infected sausage, for instance, may contain as many as one hundred thousand larval worms per ounce. A million larvae could easily be ingested during a meal.'

Shamelessly, my wife (who rather enjoys these performances) egged me on: 'Are the worms particularly dangerous?'

'Yes, indeed', I replied, with somewhat more enthusiasm than was perhaps warranted. 'The worms burrow through the intestine and travel throughout the body, boring into muscles, brains, et cetera. There is no good cure, and in Europe of the middle Ages, whole villages might become infected—many people dying horribly.'

I vaguely noticed that the gentleman across the table had stopped eating, a gefilte-fish ball poised precariously between bowl and mouth. There was no stopping me now, however, I had their attention, and I was going to keep it.

'The whole topic of parasitism is a fascinating one. Most human parasites are very well adapted to their hosts, causing few problems. Often we never know we have them. For instance, humans can have fifty-foot long tapeworms and have few or no symptoms. You know, Jewish women were once known as a primary host for *Diphyllobothrium latum*, the 'fish tapeworm'. The larvae are found in the muscles of fish, particularly pike, which is a major constituent of gefilte fish. Women would taste partially cooked gefilte fish before the larvae had been killed, thereby becoming infected.'

The man across the table slowly pushed his untouched fish aside.

My blood surged with excitement, I continued. 'However, not all tapeworms are relatively innocuous. *Echinococcus granulosus* is a species that causes very large cysts, containing ten or fifteen quarts of fluid. Humans catch the worms by swallowing their eggs, which they can get from dogs. The adult worms live in the dog's intestines, the eggs are expelled in the faeces, and the dogs often have them on their tongues, after licking their rear ends. People become infected when they allow dogs to kiss them.'

As one, all at the table turned toward the family schnauzer, who slunk off guiltily.

The host made a game effort to re-channel the conversation, remarking on how warm the room seemed to be. But I was unstoppable.

'Americans are just not used to thinking about parasites. We don't believe we have them. Actually, there are several types that commonly infect us. For instance, some of us here have an amoeba, *Entamoeba gingivalis*, in our mouths. It does not seem to do any harm, just sort of sits about on our gumlines, waiting for an occasional white blood cell to pop out.'

At the end of the table, a woman absently poured wine on the tablecloth.

'When I taught the parasitology lab', I said, savoring the memory, 'the last laboratory session was faeces day, when

336

everyone brought in their own specimens. We would find various amoebae and once in a while a worm egg. People really got into it. A woman I know had contracted amoebic dysentery in Mexico. She went off the anti-amoebic drugs she was taking, just to build up a large enough population to show the class.'

Here I was shaken from my reverie by the white faces of my companies.

The conversation soon returned (on a somewhat subdued note) to the relative merits of chicken versus beef brisket as a main dish, but for the rest of the evening the group's enthusiasm for such topics flagged.

I, on the other hand, remained jovial. I had engaged in a brilliant discourse and broadened the horizons of my companions, while never once mentioning mucus at the dinner table.

---

# Black holes

*James Munves*

From *New Yorker* 11 April 1977, p 32. Reprinted by permission, © 1977 The New Yorker Magazine, Inc.

The biggest thing in science now is black holes. Everyone is talking about them. Professor Sigmund Schmeltzer, of Göttingen, has just conducted a symposium on black holes in Aspen, Colorado. Dr T Wycliffe Meazley, of Utah A. & M., is working on black holes in Heidelberg. The junior class of the Massachusetts Institute of Technology is testing black holes in the basement of Beaumont Laboratory. Observers from Lockheed and I.B.M. are snooping around Göttingen. Dr Meazley had lunch with a vice-president of Westinghouse. Professor Pierson Flintleach, of M.I.T., is preparing a black-hole bibliography for the Pentagon. What is a black hole? Why all the interest?

A black hole is a hole that is really black. It has no color at all. It actually sucks in any light that happens to be hanging around it, including its own, which means that it

Rectangular    Grapefruit    Rectangular
black hole                   black hole
                             passing in
                             front of
                             grapefruit

Figure 1

can only be seen when it is in front of something else, like a grapefruit.

Black holes can be dangerous. If you were to run into a black hole in Central Park at 10 pm you probably wouldn't know what had happened to you. If you ran into a black hole in Central Park *no one would ever see you again*. That's why the Pentagon is interested. There are a lot of people the Pentagon never wants to see again, like Paul Warnke.

Black holes, once so rare, are becoming almost common. Under development is a rectangular black hole that can be attached to a television set to take care of commercials. (You should make certain, however, that it is properly installed.) When the pretty waitress in the commercial slides a McDonald's hamburger toward you, your rectangular black hole can be tuned to make the hamburger disappear. The great thing is, the hamburger disappears not only from your set but from every other set in the country—even sets that are not switched on. Also, the actual hamburger that was filmed six months before, in order to make the commercial, disappears, too, together with Ronald McDonald and the composer of the McDonald's jingle and the McDonald's account executive at Forbush, Scrimshaw & Nash.

There is also a toroidal, or doughnut-shaped, black hole (Figure 2) that can absorb a market-research interviewer through a keyhole or a literary agent through a Martini glass, and Professor Schmeltzer is reported to be working

| Bagel | Toroidal black hole passing in front of bagel |

**Figure 2**

on a lozenge-size black hole that, taken internally, will perform an appendectomy in less time than it takes to read this sentence. Remember, though: if your black hole is not properly installed you are in trouble. A man in Roslyn, Long Island, who acquired a trapezoidal black hole in order to eliminate a small stand of poison sumac failed to notice one loose screw, and his garden and driveway disappeared, together with his living room, the bottom half of the staircase, and the first five volumes of the Micropaedia section of his new *Encyclopaedia Britannica*.

It is difficult to believe that just a few years ago it took an eight-man team from Cornell most of a summer to produce one inverted black hole. An inverted black hole is a black hole that pierces the fabric of time. Suppose, for example, you are a suburban matron who has just written two letters, one reminding your in-laws that they are expected for Memorial Day weekend, and the other to your son's college roommate to change the site of your forthcoming assignation from the Copley Plaza to the Sheraton Boston. Suppose further that, having dropped the letters in the mailbox at 3 pm, you suddenly realize that you have put them in the wrong envelopes. With an inverted black hole, you can slip back to 2:30 pm, when you were just signing the letter to your son's roommate, and put it in the *right* envelope, or even write two new letters cancelling both of these unpromising engagements.

Not everyone is equipped to handle an inverted black hole successfully. Mr Sidney Platelet, a retired closed-end-

fund consultant, of Red Wing, Minnesota, recently purchased an inverted black hole in order to obtain
newspapers one day in advance—a scheme he thought
would help him in making personal investment decisions.
Unfortunately, when he eagerly opened the front door of
his split-level ranch at seven-fifty the following morning,
it turned out to be seven-fifty-one the *previous* morning,
and he ran into himself entering the house with the
previous day's Minneapolis *Tribune* (which he had already
read) and the milk. This was upsetting to Mr Platelet, who
was in the habit of breakfasting alone. Finding himself
admitting an additional duplicate of himself with each
passing day, Mr Platelet soon ran out of beds and, unable
to stand the squabbling over coffee and oatmeal and who
would get first crack at the paper, had to resettle himself
in Phoenix, Arizona.

Black holes never wear out or become obsolete. A black
hole has no moving parts and keeps itself clean. A black
hole is extremely useful in dealing with contemporary
crises like the power crunch and environmental pollution.
Sam Epping, the owner of Pete's Diner in Summit, New
Jersey, has recently been using an oval black hole for
waste disposal and to light up his neon sign, thus saving
on electric bills as well as on garbage service. Of course,
after five months of silently disposing of potato peelings,
coffee dregs, eggshells, cans, jars, and old copies of the
Sunday New York *Times*, Epping's black hole has
somewhat increased in size. At the time of delivery, it
fitted in a shoebox. Now it is bigger than four trailer
trucks, and Epping's customers can no longer park in the
lot behind the diner. All deliveries have to be made at the
front door, and Fred, the night short-order cook, is getting
nervous about the way chicken bones are being snatched
out of his hand.

Once a black hole is installed, it is probably there for
good, and home repairs are inadvisable. If, for example,
that householder in Roslyn now tried to adjust his faulty
black hole, his screwdriver wouldn't be the only thing
affected. Before he knew it, he might be two miles long

and thin as a needle, or one hundred and forty-six acres in area and flat as a coat of paint, or part of a clump of poison sumac—depending on whether one follows the Schmeltzer hypothesis or the most recent Meazley or Flintleach theories. It is impossible to see inside a black hole, so you can take your pick.

This brings us back to Dr Meazley. Meazley says that black holes are ushering in a new industrial revolution. 'Families will pass their black holes on to their children', Dr Meazley said recently. 'Or to their friends, if the children have been careless.' He predicts that black holes will become heirlooms and that each year more black holes will be produced to satisfy an ever-expanding market finding new uses for these versatile gadgets. Asked if the number of black holes might increase to the point where there was little room left for things, Dr Meazley replied that one can either be forward-looking and contemplate what is gained or a stick-in-the-mud and dwell on what is lost. No progress, he said, comes without some inconvenience.

Figure 3. Famous black hole of Calcutta

---

## Sonnet on a horological parasite

*By A Crostic*

RBP from Kenneth C Parkes (ed) 1983 *The Antic Alcid, An Anthology of THE AUKLET* (New York: The American Ornithologists' Union).

C  *onsider how the Cuckoo bird, my son,*
U  *pon whose family crest there lies a blot;*
C  *onnubial relations are not*
U  *nknown amongst the birds; yea, all save one*
L  *ove husband, wife and chick (and even egg)*
U  *ncommon well, and revel in the joy*
S  *upreme of parenting a girl or boy.*

C onsider, on the other hand, I beg,
A bird who avian tradition scorns,
N ot caring if her offspring doth survive
O r die. To neighbors' nests like bee to hive
R ush ovulating Cuckoos, April morns.
U nholy bird, while others tend her flock,
S he sits and calls the hours in a clock.

---

# The mathematical physicist

*Henry Petroski*

RBP from
*Southern Humanitiies Review*
**8** No 2 p 184
(1972).

*Embedded in a matrix of mistakes*
*And slips of signs, his next equation lies*
*About its symmetry. Among the lines*
*Of exercise and bold heuristic thrusts*
*Of algebra and calculus, it takes*
*His magic mirror mind to recognize*
*A juxtaposition that unifies*
*His theory of another universe.*

*Extracting the law from the accidents,*
*He calls it* Theorem *and proceeds to prove*
*It logically follows from stronger laws.*
*He makes some definitions and extends*
*The theorem more and more and marvels at the rules*
*His universe follows, effect from cause.*

---

# The hazards of WATER

RBP from Trevor A Kletz *Chem. & Eng. News* October 31, 1977, p 44. © 1977 American Chemical Society.

Daniel Morris has sent from Seattle a document he received that is alleged to have been drafted by T A Kletz, Division Safety Advisor, Imperial Chemical Industries. Excerpts follow:

'ICI has announced the discovery of a new fire-fighting agent. ... Known as WATER (Wonderful and Total

342

Extinguishing Resource), it augments, rather than replaces, existing agents such as dry powder and BCF which have been in use from time immemorial. ... Though required in large quantities, it is fairly cheap to produce and it is intended that quantities of about a million gallons should be stored in urban areas and near other installations of high risk ready for immediate use.

'ICI's new proposals are already encountering strong opposition. ... Professor Connie Barriner has pointed out that, if anyone immersed their head in a bucket of WATER, it would prove fatal in as little as three minutes. ... A spokesman from the Fire Brigades said ... it had been reported that WATER was a constituent of beer. Did this mean that firemen would be intoxicated by the fumes? The Friends of the World said that they had obtained a sample of WATER and found it caused clothes to shrink. If it did this to cotton, what would it do to men? A Local Authority spokesman said that he would strongly oppose ... a WATER reservoir in his area unless the most stringent precautions were followed. Open ponds were certainly not acceptable. What would prevent people from falling in them?

'In the House of Commons yesterday, the Home Secretary [said] ... a full investigation was needed and the Major Hazards Group would be asked to report.'

---

## The art of medicine

RBP from Samuel Sterans *Perspectives Biol. Med.* **21** No 1, Autumn 1977, p viii.

*Basic is the patient's notion*
*yours will be the magic potion.*

*Next, non-verbally convey*
*you know all there is to say.*

*Third, know how to listen well;*
*the patient knows what there's to tell.*

*Fourth, call upon an ample store*
*of ancient and of modern lore.*

*Fifth, choose every word with care;*
*never leave him in despair.*

*Finally, remember that*
*it might be you who wears his hat.*

*Comfort when you cannot cure;*
*dying's what we all endure.*

# Parody

Behavioral science: the science of pulling habits out of rats.

*Douglas Busch*

# Encounters with quantum electrodynamics

A delightful series of cartoons on quantum elec-trodynamics was originated by Robert P Bukata while a graduate student. The cartoons appeared as fillers in *Physics in Canada* circa 1962–63. The cartoons have been redrawn using new characters but the same text of Bukata's 'balloons' by Andrew Slocombe of IOP Publishing Ltd. Please find them reproduced throughout this book for your enjoyment and edification.

# Better names for units

RBP from W H Snedegar *Am. J. Phys.* **51** 8 August 1983, p 684

*Letter to the editor*

In response to the call for 'new units', I find it unappealing to continue attaching personal names to physical quantities. After all, how much momentum did Descartes or Huygens actually have? I propose rather that we rename all units in the evocative manner currently in use among the particle people, and list some suggestions below. I prefer the suggestive approach to the honor-the-famous-physicist approach for three reasons: (1) If units suggest their physical quantity, students with at least some sort of foggy idea of what a question is all about might do quite well in an essay type of response; (2) I can not remember personal names; and (3) I am not a famous physicist.

## Suggested revisions for SI units

| Physical quantity | Present name of unit | Proposed name of unit |
|---|---|---|
| Distance, displacement | 1 m | 1 far |
| Velocity, speed | 1 m/s | 1 far/s[a] = 1 jog |
| Acceleration (linear) | 1 m/s$^2$ | 1 far/s$^2$ = 1 pant |
| Mass | 1 kg | 1 lump |
| Force | 1 N | 1 shove = 1 lump far/s$^2$ |
| Work, energy | 1 J | 1 grunt = 1 shove far |
| Linear momentum, impulse | 1 kg m/s | 1 bump = 1 lump far/s |
| Angular momentum, impulse | 1 kg m$^2$/s | 1 grind = 1 bump far |
| Torque, moment of force | 1 N m | 1 twist = 1 shove far |
| Moment of inertia | 1 kg m$^2$ | 1 flab = 1 lump far$^2$ |
| Pressure | 1 pascal | 1 gasp = 1 shove/far$^2$ |
| Power | 1 W | 1 varoom = 1 grunt/s |
| Intensity level | 1 dB | 1 yell = 10 log $I/I_0$ where $I$ is in varooms/far$^2$ |

[a] I contemplated suggesting a new name for the time unit 'second', such as one 'fidget' or one 'yawn', but decided that I might not then be taken seriously.

If the scientific community should adopt these suggestions, I will be most happy to work out a similarly useful designation for physical variables in other branches of the discipline. I am sure there will be lots of help.

---

## Bizarre measures

'Office humor' offered by David Broome.

$10^{18}$ minations = 1 examination
$10^{15}$ coats = 1 petacoat
$10^{12}$ bulls = 1 terabull
$10^{9}$ lows = 1 gigalow
$10^{6}$ phones = 1 megaphone
$2 \times 10^{3}$ mockingbirds = 2 kilomockingbird
10 cards = 1 decacard
$10^{-1}$ mates = 1 decimate
$10^{-2}$ mentals = 1 centimental
$10^{-2}$ pedes = 1 centipede
$10^{-3}$ ink machines = 1 millink machine
$10^{-6}$ scopes = 1 microscope
$10^{-9}$ goats = 1 nanogoat
$10^{-9}$ nannette = 1 nanonannette
$10^{-12}$ boos = 1 picoboo
$10^{-15}$ fatales = 1 femtofatale
$10^{-18}$ boys = 1 attoboy

---

# On first looking into McConnell's *Manual*

[At Lindon's request, we sent him a copy of the *Manual of Psychological Experimentation on Planarians*. In response, he sent us a dozen poems, each modeled on a different author, each referring to flatworms and research on same.—James V McConnell, Editor, *The Worm Runner's Digest*.]

**By way of preface**
**(after Edward Lear)**

*How pleasant to know the Planarian,*
*Who has filled up such acres of print!*

349

*Some rate him a bare Turbellarian,*
  *Others bellow 'Old Triclad!' and sprint!*

*His mind is not wholly unteachable,*
  *He lives in a stream under rocks;*
*His conscience, I fear, is unreachable,*
  *He hateth electrical shocks!*

*He hath auricles, eye-spots and cilia,*
  *Though he looks merely slimy and slow;*
*Paramecium doubtless is frillier*
  *(But I own I don't properly know).*

*He dwells in a cup made for custard*
  *(With holes round the edges, we hope);*
*Raw liver he eats, without mustard;*
  *When he washes, he dare not use soap.*

*Hardly proud of his numerous progeny,*
  *His sides he will split with a laugh*
*When he loves, for (while scorning misogyny)*
  *His wife is his own better half!*

*Cry the Group, wearing nylons or flannel,*
  *When he crawls up the stem of his T,*
*'He's conditioned to choose the wrong channel,*
  *That dumb old Planarian, see?'*

*He droppeth his tails in the river,*
  *He droppeth his tails on the bank;*
*Fed on Daphnia rather than liver,*
  *He wormeth around in his tank.*

*His worth is unknown till you test him;*
  *His habits are not vegetarian;*
*Ere some cannibal kinsworm ingest him,*
  *How pleasant to know the Planarian!*

\*\*\*\*\*\*

Note: the whimsical self-portrait in verse by Edward
Lear (1812–1888) begins with

*How pleasant to know Mr Lear!*
  *Who has written such volumes of stuff!*
*Some think him ill-tempered and queer,*
  *But a few think him pleasant enough.*

## Once upon a time ...

[Milton Love recasts an old nursery tale in the light of new biological theories]

RBP from Milton Love *New Scientist* 11 February 1982 p 391. © I.P.C. Magazines.

The wolf heard that Little Red Riding Hood regularly brought goodies to her granny. He was tired of trying to chase down old caribou (high in saturated fats and gamey-tasting).

'I'm going to find me a cushier ecological niche', he declared and off he went to granny's.

Arriving at granny's house, the wolf leaped over the fence, pushed open the door and jumped into bed next to the startled old woman. Snarling through his long, white teeth, he said, 'Listen, granny, only one species can occupy the "sitting in bed getting goodies from Little Red Riding Hood" niche, and that's me. So take a hike. Here's a one way ticket on the 8:12 to Pago Pago. Be on it.'

But the old woman had battled potential competitors before and she was not easily subdued. Before he would be rid of her, the wolf would have to sign over his cattle ranch tax shelter, his strong position in copper futures and his condominium in Palm Springs. 'You can have this racket', the old woman cackled, lacing her suede Pumas and motioning her lawyers out the door. 'I couldn't have strung the kid along much longer anyway.'

The wolf had acted none too soon for he barely had time to slip into granny's nightgown and hop into bed when Red Riding Hood opened the door and sauntered in.

'Hi, granny', she said, placing her hamper of brioche, prosciutto and a very pleasant pinot chardonnay on the table. 'Hey, I'm sorry, I'm late but my moped had a flat and I had to hitch up here. I got a ride from this really fine guy who is, you know, sort of a manager for that band Amino and the Acids. Well, he isn't really their manager, but he does know the cousin of their Coke supplier so he's, you know, almost their manager. Say, granny, I don't want to be personal, but has anyone ever told you what a big nose you have? I mean, I realise, it's chic to be ethnic, but have you ever thought about some reconstructive surgery?'

'All the better to smell you with!'

'Hmm? Oh yeah, the perfume—do you like it? It's called Ms. Sexpot. It's for women ambivalent about their roles in society. I think they make it from the scent glands of Himalayan yetis. You know something else, granny? You have really big ears. I think we'd need a jackhammer to pierce your lobes. Did I tell you that Tom decided that pierced ears were really passé so instead he embedded his face in epoxy. Yes, he did. Now he's got this great big terminal smile and he says the stuff has done wonders for his complexion.'

'All the better to hear you with!'

'Granny, I hate to turn this into some sort of encounter session, but may I say what big teeth you have? Especially those pointed ones in front. Have you thought of having them capped? I think it would give your face a whole new appeal. ...'

Here the wolf threw back the bed covers and ripped off his nightgown, growling, 'All the better to eat you up with!'

Little Red Riding Hood looked at him curiously and said, 'You realise that's pretty aberrant behaviour. If you are not careful, you're going to be selected against.'

'Huh?' mumbled the wolf, puzzlement on his face.

'Well', said the girl, in somewhat pedantic tones, 'You have, I take it, outcompeted granny for this niche. Therefore, you are supposed to wait in bed while I come over five days per week (as in the contract) with various and sundry goodies. Now, by gobbling me up, you are drastically lowering the diversity of this community. There is, after all, only one Little Red Riding Hood.'

The wolf looked about, disturbed. Obviously he had not comtemplated the long-term ramifications of his intended act. 'I ... ', he stammered, 'I mean ... I didn't realise ...'

Little Red Riding Hood pressed on inexorably. 'You see, when you outcompete another species for a resource, it is your responsibility to maintain that resource. Naturally, once the word is out that you have lowered the diversity of our system, instability will set in. Think about it.' And she walked out.

The wolf was virtually paralysed with indecision. His whole purpose in occupying this niche was to gobble up Little Red Riding Hood. But as a member in good standing of his ecological community, he realised how inappropriate it would be to contribute to community instability.

Well, to make a long story short, Little Red Riding Hood did not return to her granny's house and was not eaten by the wolf. Instead she accepted a position as an *au pair* for a family of Yorkshire Terriers and was moderately happy. Granny did not fare as well, losing most of her money going short on copper futures. Probably the best off of all was the wolf. After some years of therapy, he understood that much of his problem stemmed from an inability to express his emotions, and after gobbling up his therapist, he found a measure of contentment as the manager of a small water boiling plant.

*Moral:* Competitive exclusion breeds confusion.

---

# The anti-Brownian movement

[A recently discovered manuscript appears to record some serious thoughts of Robert Browning on the physics of his day and some that was to come. It promises to add nothing whatsoever to the history and philosophy of science.]

RBP from Melbron R Mayfield *Physics Today* **26** 55 (1966) (condensed).

Will you two please understand that the Brownian Movement is not a socio – political excursion into the sordid and the uncouth. As a matter of fact, it is *Brownian* Movement, not Browning Movement. There exists a relatively unknown biologist whose name is Robert Brown and he did detect some kind of movement of particles in a liquid about the time I was 16 years old. I think this is a discovery of some value. After all, I am somewhat involved with biology myself; you know—thorns and snails and dew on the hillside and that sort of thing.

Maybe I ought to write something about this Brownian Movement that will put me in a good light regardless of what the papers say. I probably should check to see that friend Tennyson hasn't written something already. Now that he's a cabinet member, with foundation money, he's always digging around in this business of the earth moving eastward, and filling the skies with airy navies and that sort of nonsense. Let's see, what's his area code? — Oh, Al? Bob here. You heard anything about the Brownian Movement? No, my Lord, Al, it is *not* my movement. It's a scientific kick, and one of us ought to write knowledgeably about it... Oh, you'd like me to ghost this one for you—Locksley Hall style? Righto. Well, thanks Al. See you in Court.

Now—this Brownian Movement has got to move. Maybe like this...

*As I dipped into the fluid*
*All those little clods of dirt,*
*Saw them move and hit each other*
*Without seeming to be hurt,*

*As they frittered on the surface*
*Going every which way,*
*Fast and slow, but always random,*
*All those tiny bits of clay,*

*Came to me a mighty wisdom,*
*Far beyond the normal size,*
*And I heard in tones of thunder*
*'This is worth a Nobel Prize.'*

*What should be my next achievement?*
*Shall I seek out more renown?*

*Find a fuller explanation*
*Of this movement, name of Brown?*

*I will ponder on the makeup*
*Of these tiny moving things.*
*All are small but some are smaller?*
*Whirling round in busy rings?*

*Big ones plus and others minus?*
*Some without a charge at all?*
*Protons, neutrons and electrons?*
*Nice new names for me to call.*

*These should be the fundamental*
*Parts that make up our universe.*
*Still I feel there must be others*
*(Even though I feel in verse.)*

*Up to now all my electrons*
*Have been minus—all alike—*
*Now I've got a plus electron;*
*What to call this lucky strike?*

*Antiprotons, antineutrons*
*Also live somewhere in space,*
*Merging into antimatter*
*Never leaving any trace.*

*Antiplanets need these forces,*
*In an antiearthy way,*
*For the antifolks who live there*
*Antiday by antiday.*

*Here the baryons and mesons,*
*Superlarge, but very strange,*
*Group themselves in odd arrangement,*
*Energized in such a range.*

*Thus my patient observations*
*And my theories neatly drawn*
*Could have brought a science noontime*
*Long before the science dawn.*

*Could have—but it didn't happen,*
*Though my records all are clear:*
*Photographs and antidata,*
*And I have them all right here.*

*What deterred me from the glory*
*That I earned before my time?*
*Antiscience Neilsen ratings*
*Cut me off without a dime.*

*Left me dead without a sponsor,*
*Shut out by a soccer game.*
*Left me just my notes and pictures*
*And a share of antifame.*

Oh, well, as a great poet will probably say, 'It's better to have loved...' And to be very honest, I feel good about the Brownian Movement, but not about the anti-Brownian Movement. Well, the whole anti-situation—all fifteen antiparticles, antimatter, antigalaxies, antiwhatever—if these really do exist, then what the antiheaven for?

---

# Body ritual among the Nacirema

*Horace Miner*

RBP of the American Anthropological Association from *American Anthropologist* **58** 3, June 1956.

The anthropologist has become so familiar with the diversity of ways in which different peoples behave in similar situations that he is not apt to be surprised by even the most exotic customs. In fact, if all of the logically possible combinations of behavior have not been found somewhere in the world, he is apt to suspect that they must be present in some yet undescribed tribe. This point has, in fact, been expressed with respect to clan organization by Murdock (1949:71). In this light, the magical beliefs and practices of the Nacirema present such unusual aspects that it seems desirable to describe them as an example of the extremes to which human behavior can go.

Professor Linton first brought the ritual of the Nacirema to the attention of anthropologists twenty years ago (1936:326), but the culture of this people is still very poorly understood. They are a North American group living in the territory between the Canadian Cree, the Yaqui and Tarahumare of Mexico, and the Carib and Arawak of the Antilles. Little is known of their origin, although tradition states that they come from the east. According to Nacirema mythology, their nation was originated by a culture hero, Notgnihsaw, who is otherwise known for two great feats of strength—the throwing of a piece of wampum across the river Pa-To-Mac and the chopping down of a cherry tree in which the Spirit of Truth resided.

Nacirema culture is characterized by a highly developed market economy which has evolved in a rich natural habitat. While much of the people's time is devoted to economic pursuits, a large part of the fruits of these labors and a considerable portion of the day are spent in ritual activity. The focus of this activity is the human body, the appearance and health of which loom as a dominant concern in the ethos of the people. While such a concern is certainly not unusual, its ceremonial aspects and associated philosophy are unique.

The fundamental belief underlying the whole system appears to be that the human body is ugly and that its natural tendency is to debility and disease. Incarcerated in such a body, man's only hope is to avert these characteristics through the use of the powerful influences of ritual and ceremony. Every household has one or more shrines devoted to this purpose. The more powerful individuals in the society have several shrines in their houses and, in fact, the opulence of a house is often referred to in terms of the number of such ritual centers it possesses. Most houses are of wattle and daub construction, but the shrine rooms of the more wealthy are walled with stone. Poorer families imitate the rich by applying pottery plaques to their shrine walls.

While each family has at least one such shrine, the rituals associated with it are not family ceremonies but are private and secret. The rites are normally only discussed with children, and then only during the period when they are being initiated into these mysteries. I was able, however, to establish sufficient rapport with the natives to examine these shrines and to have the rituals described to me.

The focal point of the shrine is a box or chest which is built into the wall. In this chest are kept the many charms and magical potions without which no native believes he could live. These preparations are secured from a variety of specialized practitioners. The most powerful of these are the medicine men, whose assistance must be rewarded with substantial gifts. However, the medicine men do not provide the curative potions for their clients, but decide what the ingredients should be and then write them down in an ancient and secret language. This writing is understood only by the medicine men and by the herbalists who, for another gift, provide the required charm.

The charm is not disposed of after it has served its purpose, but is placed in the charm-box of the household shrine. As these magical materials are specific for certain ills, and the real or imagined maladies of the people are many, the charm-box is usually full to overflowing. The magical packets are so numerous that people forget what their purposes were and fear to use them again. While the natives are very vague on this point, we can only assume that the idea in retaining all the old magical materials is that their presence in the charm-box, before which the body rituals are conducted, will in some way protect the worshipper.

Beneath the charm-box is a small font. Each day every member of the family, in succession, enters the shrine room, bows his head before the charm-box, mingles different sorts of holy water in the font, and proceeds with a brief rite of ablution. The holy waters are secured from the Water Temple of the community, where the priests conduct elaborate ceremonies to make the liquid ritually pure.

In the hierarchy of magical practitioners, and below the

medicine men in prestige, are specialists whose designation is best translated 'holy-mouth-men'. The Nacirema have an almost pathological horror of and fascination with the mouth, the condition of which is believed to have a supernatural influence on all social relationships. Were it not for the rituals of the mouth, they believe that their teeth would fall out, their gums bleed, their jaws shrink, their friends desert them, and their lovers reject them. They also believe that a strong relationship exists between oral and moral characteristics. For example, there is a ritual ablution of the mouth for children which is supposed to improve their moral fiber.

The daily body ritual performed by everyone includes a mouth-rite. Despite the fact that these people are so punctilious about care of the mouth, this rite involves a practice which strikes the uninitiated stranger as revolting. It was reported to me that the ritual consists of inserting a small bundle of hog hairs into the mouth, along with certain magical powders, and then moving the bundle in a highly formalized series of gestures.

In addition to the private mouth-rite, the people seek out a holy-mouth-man once or twice a year. These practitioners have an impressive set of paraphernalia, consisting of a variety of augers, awls, probes, and prods. The use of these objects in the exorcism of the evils of the mouth involves almost unbelievable ritual torture of the client. The holy-mouth-man opens the client's mouth and, using the above mentioned tools, enlarges any holes which decay may have created in the teeth. Magical materials are put into these holes. If there are no naturally occurring holes in the teeth, large sections of one or more teeth are gouged out so that the supernatural substance can be applied. In the client's view, the purpose of these ministrations is to arrest decay and to draw friends. The extremely sacred and traditional character of the rite is evident in the fact that the natives return to the holy-mouth-men year after year, despite the fact that their teeth continue to decay.

It is to be hoped that, when a thorough study of the

Nacirema is made, there will be careful inquiry into the personality structure of these people. One has but to watch the gleam in the eye of a holy-mouth-man, as he jabs an awl into an exposed nerve, to suspect that a certain amount of sadism is involved. If this can be established, a very interesting pattern emerges, for most of the population shows definite masochistic tendencies. It was to these that Professor Linton referred in discussing a distinctive part of the daily body ritual which is performed only by men. This part of the rite involves scraping and lacerating the surface of the face with a sharp instrument. Special women's rites are performed only four times during each lunar month, but what they lack in frequency is made up in barbarity. As part of this ceremony, women bake their heads in small ovens for about an hour. The theoretically interesting point is that what seems to be a preponderantly masochistic people have developed sadistic specialists.

The medicine men have an imposing temple, or *latipso*, in every community of any size. The more elaborate ceremonies required to treat very sick patients can only be performed at this temple. These ceremonies involve not only the thaumaturge but a permanent group of vestal maidens who move sedately about the temple chambers in distinctive costume and headdress.

The *latipso* ceremonies are so harsh that it is phenomenal that a fair proportion of the really sick natives who enter the temple ever recover. Small children whose indoctrination is still incomplete have been known to resist attempts to take them to the temple because 'that is where you go to die'. Despite this fact, sick adults are not only willing but eager to undergo the protracted ritual purification, if they can afford to do so. No matter how ill the supplicant or how grave the emergency, the guardians of many temples will not admit a client if he cannot give a rich gift to the custodian. Even after one has gained admission and survived the ceremonies, the guardians will not permit the neophyte to leave until he makes still another gift.

The supplicant entering the temple is first stripped of all

his or her clothes. In every-day life the Nacirema avoids exposure of his body and its natural functions. Bathing and excretory acts are performed only in the secrecy of the household shrine, where they are ritualized as part of the body-rites. Psychological shock results from the fact that body secrecy is suddenly lost upon entry into the *latipso*. A man, whose own wife has never seen him in an excretory act, suddenly finds himself naked and assisted by a vestal maiden while he performs his natural functions into a sacred vessel. This sort of ceremonial treatment is necessitated by the fact that the excreta are used by a diviner to ascertain the course and nature of the client's sickness. Female clients, on the other hand, find their naked bodies are subjected to the scrutiny, manipulation and prodding of the medicine men.

Few supplicants in the temple are well enough to do anything but lie on their hard beds. The daily ceremonies, like the rites of the holy-mouth-men, involve discomfort and torture. With ritual precision, the vestals awaken their miserable charges each dawn and roll them about on their beds of pain while performing ablutions, in the formal movements of which the maidens are highly trained. At other times they insert magic wands in the supplicant's mouth or force him to eat substances which are supposed to be healing. From time to time the medicine men come to their clients and jab magically treated needles into their flesh. The fact that these temple ceremonies may not cure, and may even kill the neophyte, in no way decreases the people's faith in the medicine men.

There remains one other kind of practitioner, known as a 'listener'. This witch-doctor has the power to exorcise the devils that lodge in the heads of people who have been bewitched. The Nacirema believe that parents bewitch their own children. Mothers are particularly suspected of putting a curse on children while teaching them the secret body rituals. The counter-magic of the witch-doctor is unusual in its lack of ritual. The patient simply tells the 'listener' all his troubles and fears, beginning with the earliest difficulties he can remember. The memory displayed by the Nacirema in these exorcism sessions is truly remarkable. It is not uncommon for the patient to bemoan the rejection he felt upon being weaned as a babe, and a few individuals even see their troubles going back to the traumatic effects of their own birth.

In conclusion, mention must be made of certain practices which have their base in native esthetics but which depend upon the pervasive aversion to the natural body and its functions. There are ritual fasts to make fat people thin and ceremonial feasts to make thin people fat. Still other rites are used to make women's breasts larger if they are small, and smaller if they are large. General dissatisfaction with breast shape is symbolized in the fact that the ideal form is virtually outside the range of human variation. A few women afflicted with almost inhuman hypermammary development are so idolized that they make a handsome living by simply going from village to village and permitting the natives to stare at them for a fee.

Reference has already been made to the fact that excretory functions are ritualized, routinized, and relegated to secrecy. Natural reproductive functions are similarly distorted. Intercourse is taboo as a topic and scheduled as an act. Efforts are made to avoid pregnancy by the use of magical materials or by limiting intercourse to certain phases of the moon. Conception is actually very infrequent. When pregnant, women dress so as to hide their conditions. Parturition takes place in secret, without friends or relatives to assist, and the majority of women do not nurse their infants.

Our review of the ritual life of the Nacirema has certainly shown them to be a magic-ridden people. It is hard to understand how they managed to exist so long under the burdens which they have imposed upon themselves. But even such exotic customs as these take on real meaning when they are viewed with the insight provided by Malinowski when he wrote (1948:70):

Looking from far and above, from our high places

362

of safety in the developed civilization, it is easy to see all the crudity and irrelevance of magic. But without its power and guidance early man could not have mastered his practical difficulties as he has done, nor could man have advanced to the higher stages of civilization.

### References

Linton Ralph 1936 *The Study of Man* (New York: D. Appleton-Century Co)

Malinowski Bronislaw 1948 *Magic, Science, and Religion* (Glencoe: The Free Press)

Murdock George P 1949 *Social Structure* (New York: The Macmillan Co).

---

# The revolution of the corpuscle

[The contrasting viewpoints as between the electromagnetic nature of light and the quantum nature are set forth in the following three stanzas.]

From A A Robb (1926) in *Post-Prandial Proceedings of the Cavendish Society* **6** (Cambridge: Bowes & Bowes).

*Air*: 'The Interfering Parrot'. (*The Geisha.*)

1  *A corpuscle once did oscillate too quickly to and fro,*
   *He always raised disturbances wherever he did go.*
   *He struggled hard for freedom against a powerful foe—*
   *An atom—who would not let him go.*
   *The aether trembled at his agitations*
   *In a manner so familiar that I only need say,*
   *In accordance with Clerk Maxwell's six equations*
   *It tickled people's optics far away.*
      *You can feel the way it's done,*
      *You may trace them as they run—*
   d$\gamma$ *by* d*y less* d$\beta$ *by* d*z is equal* $K \cdot$ d*X*/d*t.*
      ...
   *While the curl of (X, Y, Z) is the minus* d/d*t of the*
   *vector (a, b, c).*

2  *Some professional agitators only holler till they're hoarse.*
   *But this plucky little corpuscle pursued another course,*
   *And finally resorted to electromotive force,*

*Resorted to electromotive force.*
*The medium quaked in dread anticipation,*
*It feared that its equations might be somewhat too*
  *abstruse,*
*And not admit of finite integration*
*In case the little corpuscle got loose.*
  *For there was a lot of gas*
  *Through which he had to pass,*
  *And in case he was too rash,*
  *There was sure to be a smash,*
  *Resulting in a flash.*
*Then* d$\gamma$ *by* dy *less* d$\beta$ *by* dz
  *would equal* K · dX/dt.
  *...*
*While the curl of (X, Y, Z) would be minus* d/dt *of the*
  *vector (a, b, c).*

3  *The corpuscle radiated until he had conceived*
  *A plan by which his freedom might be easily achieved;*
  *I'll not go into details for I might not be believed,*
  *Indeed, I'm sure I should not be believed.*
  *However, there was one decisive action,*
  *The atom and the corpuscle each made a single charge,*
  *But the atom could not hold him in subjection,*
  *Though something like a thousand times as large.*

# Spoof

I once asked an American scientist in 'defense biology' what they were looking for. He said, 'A cure for metabolism'.

# The Schuss-yucca
## (*Yucca Whipplei*, var. *Schuss*)

*Gustav Albrecht*

RBP from *The Scientific Monthly* **75**, October 1952 pp 250–2. Copyright 1952 by the AAAS.

One of the most amazing, and still unexplained, phenomena in botany is the *Schuss-yucca*, a rare variety of the chaparral yucca (*Yucca Whipplei*), which occurs here and there about Chilao Flat in the San Gabriel Mountains north of Pasadena, a locale noted for its queer flora and fauna. The normal variety of Yucca Whipplei grows for many years as a hemisphere of sharp and awesome spines (Spanish bayonet); then, some spring day, a large shoot rises ten to twenty fee in a period of two or three weeks. blooms, and dies. The Schuss-yucca of Chilao Flat does this in a matter of minutes or even seconds! (See photographs.)

Although described in a brief note in Liebig's *Annalen* (1853), the first thorough investigation of this amazing plant was made by the eminent German botanist Professor Ferdinand Grünspann, who visited Chilao Flat, riding a burro, in the spring of 1890, and who devoted Volumes 13 and 14 of his exhaustive twenty-volume work, *Handbuch der Yucca* (Leipzig: Schmutzig-verlag (1893)—now out of print and hard to find), to a description of this remarkable variety of the chaparral yucca. In spite of Professor Grünspann's reputation and careful research, his observations on the Schuss-yucca, (a name he himself applied to it from the German *Schuss*, 'to shoot up') were not credited by contemporary botanists, who considered it a hoax (*zum lachen*). Although most of the professor's work was undoubtedly accurate, he apparently let himself be taken in by certain tall stories of the Indians, one tale in particular doing much to discredit his whole study. Grünspann tells of the Spanish desperado and cattle rustler Vasquez being impaled in midair by a Schuss-yucca while he (Vasquez) was jumping over the plant. This story possesses undoubted charm, but it is probably a canard. The Schuss-yucca does shoot up with amazing rapidity in a matter of seconds, but the shoot is soft, like a giant asparagus, and is frequently eaten by deer (which can be seen waiting about near plants that they somehow

366

know are ready to sprout); hence, it could not reasonably be expected to harm a full-grown Spaniard.

This story, and the lack of imagination common at that time, combined to discredit Professor Grünspann. It is said that the professor's dying words were '*Es schiesst doch!*'—reminiscent of Galileo's famous remark some centuries earlier, '*Eppur si muove!*' But rumors about the Schuss-yucca, which the natives picturesquely refer to as the 'jumpin' Yuccy', have persisted about Chilao Flat. One rumor originated at Mount Wilson a few miles to the south from an astronomer who was fooling around with the 100-inch telescope between exposures of the planet Mercury, and another from a skier who had got off his course in a *slalom* race down Mount Waterman.

Since I had myself seen and photographed another queer yucca in this area last year—*Yucca Whipplei*, var. *bifurcata* (mihi)—*Science* **115** 219 (1952)), I determined to investigate the Schuss-yucca, and if possible bring back incontrovertible photographic evidence.

My model and I, laden with cameras and a fifth of antivenin (the rattlesnakes, too, are said to be unusually swift and accurate at Chilao Flat), haunted the region for days. We were guided by the deer, whose uncanny instinct tells them which yucca plants to watch, and we were finally rewarded for our perseverance, as the accompanying photographs show. I used an automatic Rolleiflex, taking the photographs at one-second intervals, and although the exposures were only 1/100 of a second, there is little blurring of the fast-moving sprout. The amazement on the face of the model, who was somewhat dubious of the whole affair, is clearly evident.

It is somewhat sad that cameras were not as well developed in Professor Grünspann's time—he might have died a happier man. It is also regrettable that scientists as a class were so skeptical then. Scientists today realize that anything is possible, whether Schuss-yuccas or extraterrestrial flying saucers—particularly when reported by trained and reliable observers and accompanied by good photographs.

# When the data nets laughed

*Simplicius (James Dyson)*

RBP from the National Physical Laboratory. *NPL News* No 310, 21 February 1976. Crown Copyright.

Ed Perlmutter looked out of the window at the Arecibo thousand-foot radio dish. 'Yeah, Hank—very efficient. No wonder time on it is too expensive for the sort of bread NSF gives us. Can't we get better value for the few minutes we can afford?'

'Well, Ed, I've had an idea. Can't we receive them all at once—$H_2$, OH, NO, HCN, HCHO—and sort them out by computer? Sure, with such a broad band, the signal-to-noise'll be lousy, but say we use a filter that passes narrow bands at all those frequencies, and nothing from the continuum in between? We'd get rid of most of the noise that way.'

369

'Sure, Hank, great idea. I'll get the computer to design one.' And Ed turned to the computer terminal to key in the problem.

At Jodrell Bank, Roy Sattersthwaite looked out of the window at the big Mk 2 dish and sighed. 'A very able collector of energy, Bill, but that doesn't help us any. That star-breeding-ground in Orion is a devil's brew of chemicals—$H_2$, OH, NO, HCN, HCHO—and all of them superradiant, so that you can hardly detect the continuum; and, as we want to measure the temperature, the continuum is all we're interested in.'

'Well, there's a way round that', said Bill. 'If we use a filter that stops narrow bands just at those frequencies, we shall receive only the continuum, and the receiver won't be overloaded by the emission lines. I'm sure the computer can design us a filter like that.'

'Jolly good idea', said Roy, and began to tap out the problem on the terminal.

In a few minutes the computer began to draw out the transmission function of the filter it had designed on the print-out. It was an almost horizontal line, save for sharp dips almost to zero at points corresponding to the emission lines of the offending molecules. Roy and Bill watched with satisfaction as the tracing was completed. Roy said 'Well, I think that ought to do it. Now we have to translate that into hardware, but that shouldn't ...' His voice trailed off into silence. The computer was beginning to draw out another curve which was almost the mirror-image of the first one, with pass-bands instead of stop-bands at the emission lines. It was, in fact, the characteristic curve of the filter required by Ed and Hank in Arecibo.

There was a short silence. Then the computer began to print-out alphanumeric characters apparently randomly at immense speed. Paper spouted from the machine for ten seconds, then silence. Ten seconds later the action started again, and the sequence of alternate frenzied print-out and inaction continued.

Roy telephoned a technician at the main University site,

who said 'Every print-out in the University is doing it. Listen!' And over the telephone Roy could hear the rhythmic storming of the automatic typewriters in Manchester. As the day wore on, it became evident that every major computer system in the world was incapacitated by the same irrational behaviour. It was six hours before the pathological behaviour died away.

Roy and Bill sat listening to the last spasmodic activity of the typewriters. Roy said 'I think I know what has happened. Virtually all the big computers in the world are tied together by land-lines or satellite, and the complexity is such that we might expect entirely new kinds of behaviour. Perhaps even behaviour of a human nature. And I believe that the system has acquired a sense of humour. You see, one source of humour is the recognition of new relationships, like the "double entendre" or the "literal" in newpaper headlines. The internal connexions in the system are good enough for it to know about both our own filter problem and Arecibo's. So, when it saw two filter characteristics which were exact mirror-images of each other, the computer found it irresistibly funny. Laughter is an activity without obvious purpose which is compulsive enough to supress all other activity, and the computer analogue of that is the printing of nonsense. Unfortunately, the result is that every large computer in the world is put out of action whenever any one of them sees something funny. And—this has only just struck me—the computers may begin to make their own jokes!'

Bill paled a little. 'My God—think of it! The digital pun!'

---

# Trans-Canada high vacuum line

[This is the proposal submitted over unreasonable facsimiles of the signatures of several prominent Canadians in December 1963 for the construction of the Trans-Canada high vacuum line. This concept arose out of an extremely contentious issue in Canadian politics, the Trans-Canada Pipeline. The stomy debates on this in the Canadian Parliament in 1956 led directly to the fall of the Liberal Government and its replacement by the Conservatives under John Diefenbaker.]

## A Neglected Project

RBP from *A Neglected Project* by L G Harrison *Science Forum* No 1 1968; *Encore le Vacuoduc* by P A Giguère *Science Forum* No 3 1968.

Sir:

I am writing to protest the apathetic attitude which has prevailed for nearly four years regarding a project which would make Canada unique among scientific nations, and which, of all scientific projects proposed in this country, would do the most to further the interests of Canadian Confederation, which stands today in such peril.

Your enlightened readership must surely know at once to what I am referring. It is of course the proposal, submitted to NRC as long ago as December 1963 by prominent representatives of certain western Universities, for a major equipment grant for the construction of a Trans-Canada High Vacuum Line. The project was very simple in concept, and envisaged the establishment of a completely vacuous link between all major Canadian universities by means of a pipe running along a great circle route from Vancouver to Montreal, with branches to all universities, the whole system being evacuated to $10^{-12}$ mm Hg. This pressure was chosen so that gas molecules in the pipe would have a mean free path exceeding 2500 miles, in order that a molecular beam might be projected from Vancouver to Montreal with the most precise velocity selection known to modern technology, since a molecule could only get from one end of the pipe to the other if it was in orbit with a velocity of $8 \times 10^5$ cm/sec with a tolerance of 0.02 cm/sec (for a one-foot diameter pipe).

The original proposal outlined fully all the difficulties which might be encountered, and the means which were proposed to overcome them. For example, the orbital molecular velocity mentioned corresponds to a temperature of about 100,000°K, and the authors of the proposal suggested generating the beam by a pulse technique involving the successive detonation of a series of ther-

372

monuclear devices, thus providing a remarkable peaceful use for nuclear weaponry. How can the federal authorities so long ignore a project which contains such a fusion of space-age thinking, nuclear technology and the familiar vacuum apparatus so dear to the heart of every physical chemist, and which would further enable every Canadian university to claim that it had the world's largest piece of equipment?

And all of this, sir, with its glass-lined pipe having a fine gold wire stretched along its axis for flash-evaporation of a clean gold coating; its trans-continental railway carrying the leak-testing equipment, and thus bringing a large labour force into the Far North through which the great circle route passes; its potential in times of emergency for alternate use as an aqueduct to divert the waters of the Columbia River into the Gulf of St. Lawrence; all this for an estimated cost of only $5 billion to the Canadian taxpayer.

Above all, sir, what excuse can there be for delay on this project in face of the overwhelming enthusiasm shown in this statement from our compatriots in French Canada, which formed part of the original submission: 'Parce qu'il faut établir entre les deux nations du Canada le ''Rapport Absolument Vide Interprovincial (RAVI)'', les Canadiens de langue française seront heureux de construire leur parti de la ligne; cependant, il faut que la ligne exprime la caractère nationale de la province. Ainsi, le fil métallique au centre de la ligne dans la province de Québec ne sera pas construit de l'or, mais de l'Argent des Créditistes.' Surely, sir, if this project were activated, we should soon hear the new rallying-call of 'Vive le Vide! Vivent les molécules libres!' going up all over la belle province.

*L G Harrison*
Department of Chemistry
University of British Columbia

\*\*\*\*\*\*

Professor P A Giguere, of Laval University, in the third issue of *Science Forum* published a letter in which he applauded Harrison's suggestion for a Trans-Canada Vacuum Line. He suggested building a tower into the stratosphere to provide a direct connection to a source of high vacuum and avoid the need for a pumping system. Regarding this project, he happily viewed the prospect of a 'country united by a vacuum'.

# Introduction to the Number 2-B Regrettor. *A 1937 engineering parody that foreshadowed thinking machines*

*E A Weiss*

Reproduced from *Annals of the History of Computing* vol 7 No 2, April 1985 pp 167–76 (abridged).

The pompous formality and rigid stuffiness of official documents often tempt humorists to write parodies, but rarely do engineering parodies have either long life or wide circulation. The anonymous underground paper, 'Number 2-B Regrettor, Description and Procedures, For People Who Think They Think', is an exception. Although written more than 45 years ago, it was liberally copied at a time when copying was far more costly than it is today. It was circulated informally through the engineering parts of the telephone, radio, audio, and electronics industries for years.

Many now-elderly electrical engineers affectionately recall the 2-B Regrettor and keep copies of its description in their nostalgia files.† The quality of its humor can be appreciated only by reading it, and its full impact will be felt only by those who knew and were subject to the conditions of its origin.

I first saw the 2-B Regrettor paper (probably a copy of the second edition) in early 1941 and encountered the third edition on a visit to MIT, probably in 1944.

†MIT professor emeritus Truman Gray, whose copy of the paper we reprint here, asked us to 'return my original so that I may refile it where it has been for all these years—under "C" for "Crackpots".'

374

My serious search started in 1983 when I realized that the paper described a forerunner of the computer and that the *Annals of the History of Computing* might be willing to reprint it.

Some respondents sent copies of the paper (chiefly the second edition), many reported having had and lost copies, and a few had never actually seen the paper but had heard of it. Through further correspondence with those who had the most information, I learned about the two authors, James M Henry and Ernest R Moore, and about the conception and formulation of the 2-B Regrettor paper.

The paper takes the form and appearance of a Bell System Practice, a standard document that went into the field with all apparatus and equipment† devised and issued by subsidiaries of the American Telephone and Telegraph Company—chiefly the operating companies, Bell Laboratories and Western Electric. In keeping with the model it parodies, the paper includes the setup, operating and calibrating instructions, and tests for the Regrettor. 'The assignment generally used to determine that the set is functioning properly is essentially that of regret at having purchased the instrument.' It spoofs the mathematics of the principles of operation, consistently mimicking the style and language of a Bell System Practice. Performance characteristics are mockingly charted, and a sketch shows the Regrettor's ease of operation in the hands of six frantic engineers.

The idea that a machine can 'regret' (and with its connections reversed 'anticipate') marks this paper as a representation of the speculations about machine intelligence in the minds of thoughtful and creative electrical engineers just before World War II.

---

†Generally (but not always), Bell System parts and components were 'apparatus', while finished assemblies were 'equipment'. The Regrettor would be classified as equipment, and a dynamically stabilized microsynchronous effort bender would be apparatus.

## Notes and comments

The following remarks explain and comment on some references in the paper that may be obscure. The decimal numbers refer to the paragraphs of the facsimile paper.

1.11. Because electronic equipment was large and bulky, it was conventionally described as being light and compact. Experienced readers would discount this language and turn directly to Subsection 2-G, which gives the true size and weight under the heading 'Mechanical Arrangement'.

2.17.a. The mention of the Republican Party refers to Democrat Franklin D Roosevelt's 1936 landslide victory over Alfred Landon, who carried only Maine and Vermont, giving Roosevelt his second term.

2.17b. The reference to Parliament and 'the British coronary incumbent' has to do with the 1937 abdication of Edward VIII.

2.19. 'Non-rectifying type meter' refers to audio-frequency panel meters that used copper oxide rectifiers.

2.21. The voltage provided to the grid, one of the elements of a vacuum tube, was called its 'bias'.

2.25. The 221-C was a real Western Electric portable radiotelephone set in the 2-mhz range. A 'D' number identifies a prototype.

4.13. Vacuum-tube grids were connected to their emitting elements through high resistances called 'grid leak resistors'.

4.15.d. A major Bell System manufacturing plant, a part of Western Electric, was in Kearny, NJ.

5.11. Income tax day was then March 15, not April 15.

5.11.b. The 1936 Democratic presidential platform had promised a reduction in the income tax.

5.11.d. 'We're With You, Mr President' was a 1936 Democratic campaign song.

6.11. This is a spoof of a characteristic mathematics section of a Bell System Practice. FDR = Franklin Delano Roosevelt; PWA, WPA, NRA, F.P.C. were federal agencies. The administration was at odds with the chief justice of the United States, Charles Evans Hughes, and hoped to cut him off. James Farley, postmaster general, had been Roosevelt's campaign manager. Haig and Haig whiskey is associated here with tippling and putting the quart before the hearse. P T Barnum was the nineteenth-century circus owner who is reputed to have said, 'There is a sucker born every minute'.

## The authors

James M Henry and Ernest R Moore were a pair of electrical engineers in the General Engineering Department of the New England Telephone and Telegraph Company in Boston. True to his French background, Jim Henry was active and volatile and is recalled as being the more witty of the two. Ernie Moore was slow in speech and movement; some thought him a bit dull. Both had wonderful senses of humor, however—Jim's sharp, Ernie's droll and dry.

Moore, about 15 years older than Henry, was a radio engineer who had been with NET&T since 1912 and was then involved in transmission and inductive coordination. His peers rated him an excellent engineer. Moore remained with NET&T until he retired. He has since died.

Henry, the principal author, was about 32 when he and Moore wrote the paper in 1937. Originally from Louisiana, he had been with the Apparatus Department of the Bell Telephone Laboratories for almost 10 years and most recently had been concerned with the development of ship-to-shore radio systems. During this project he was transferred to NET&T 'as that company was badly in need of an engineer with radio experience'. Professionally Jim Henry was a very sharp character; his ready wit could erupt at any time, often breaking the tension in a serious meeting.

During World War II Henry was a supervisor at the Naval Radar School at MIT, and he then returned to his job at NET&T. In 1951, feeling a need for more challenge, he left the Bell system. As he said in a 1964 letter to a friend, 'Most of the engineering was becoming "handbook" in form and I wanted to exercise creativity on my own'. First he joined an instrument company where he designed toroidal coil winders and managed sales. Later he moved to Trippitt Advertising, where he dreamed up advertising copy and layouts for industrial electronics products. While there, he moonlighted a proposal for a complete telecommunications system for Laos, Thailand, and South Vietnam, which was accepted by a government agency and for which he was hired as project manager. He assembled a staff for a study and prepared the report and recommendations for the system. For personal reasons he left the project and joined an MIT educational program to improve the teaching of science in U.S. high schools. This project became a separate corporation called Educational Services Inc. Henry died on January 20, 1966, and was survived by his wife, Connie, three adult children, and at least one grandchild. His wife died a few years later.

## Writing the paper

Henry stayed at the same boarding house in Morristown, NJ, as William W H Doherty, who was working on the development of high-power radiotelephone transmitters for the trans-Atlantic and broadcasting services at the Bell Labs Whippany, NJ, lab a few miles away.†

Doherty recalls the circumstances that probably laid the groundwork for the paper. 'We used to dine together evenings. He was intrigued by some of the tricky things we were using involving a time delay for processing speech signals (e.g. for "noiseless recording" and for permitting two-way conversation on our single-channel transatlantic radiotelephone) and had heard someone talking jokingly about a "1-A Anticipator". It seemed to him that if we had an anticipator we should also have a regret-

---

†Bell Labs was the development arm of Western Electric, which was a leading supplier of radio transmitting and broadcast equipment.

tor. Jim was just the kind of funster who would go ahead and incorporate this into a parody on a Bell System ''spec''.'

BULL SYSTEM PRACTICES
Special Engineering
Extraordinary Services

SECTION Z999, 998
Issue 1, 4-26-37
Provisional Standard

## NUMBER 2-B REGRETTOR

DESCRIPTION AND PROCEDURES
For People Who Think They Think
*James M Henry and Ernest R Moore*

### 1. General

**1.11** The No. 2-B Regrettor is a light weight, compact unit, recently developed specifically for use by persons whose capacities for regretting are below normal, or who, due to various activities have more to regret than can be conveniently handled without artificial aid.

**1.12** Through the use of this device it is possible for the user to have his bad moments regretted for him, and he is meanwhile left free to engage in activities which may be regretted later.

**1.13** This device may be absolutely depended upon to faithfully regret in accordance with the wishes of even the most talented of bunglers. It is guaranteed to bend every effort to perform its duties, being equipped with a dynamically stabilized microsynchronous effort bender of the cantilever type.

**1.14** By means of a simple change-over switch, the No. 2-B Regrettor can be made to rue. Days are the least difficult of all items to rue. To ''rue the day'' it is necessary merely to set the machine up in accordance with the simple instructions given under OPERATION.

a. In case there is doubt as to the simplicity of the instructions, see Part 4.

**1.15** Provision has been made for the use of the No. D-9445 Anticipatory Converter Attachment in connection with the No. 2-B Regrettor. The combined action of these two units operating in conjunction is such that the Regrettor functions in a negative sense. The net result is essentially that of an Anticipator. The advantages attendant upon such flexibility hardly require pointing out.

a. This singular effect is explained in Part 3 under CONVERTER ATTACHMENT.

**1.16** The No. 2-B Regrettor may be adjusted to give down pangs of regret if such are desired. All frequencies up to 20 pangs per second are obtainable under control of the operator. When the Pang Frequency Control is set on 20 p.p.s. the output from jacks J-10 and J-11 labelled FACETIOUS OUTPUT may be used as ringing current on the drop side of all loops within reach, if desired.

Note: This is what would be known as a Regrettable Incident.

**1.17** The No. 2-B Regrettor is designed to fit the No. 221-C Mounting.

a. This mounting is shock proof and will withstand the impact of the housing of the Regrettor which sometimes produces violent motion when suffering pang regret of high amplitude. Helical spring pang suppressors suitably placed about the frame of the mounting aid in reducing the acoustic shock to a minimum, and serve to protect the tubes from damage.

## 2. Characteristics

A. *Impedance Range*

**2.11** The No. 2-B Regrettor has been designed to satisfactorily match a wide range of emotional impedance, the actual values of which depend in large measure,

of course, upon the temperament coefficient of the user. Where this factor is known to be substantially consistent, the instrument may be relied upon to adequately perform regrets for long periods of time without re-calibration.

a. Human nature being what it is, however, it is recommended that the instrument be checked at frequent intervals in accordance with the instructions given in part 5 of this Section under CALIBRATION.

b. On the other hand perhaps not.

**2.12** A chart of the output impedance is given in figure 1.

B. *Response*

**2.13** The response of the No. 2-B Regrettor is substantially uniform over the entire sad gamut of emotional range being less than 2 db down at the perfunctory regret end and 3 db down at the black despair end.

a. Note: Where a flatter characteristic is desired the Regrettor may be flattered with a 22 type False Praise Lavisher.

1. *Caution: If an A.C. operated 22-C Lavisher is used a filter will be required to smooth out the output.*

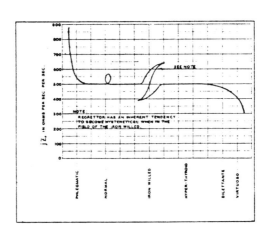

**Figure 1.** Output Impedance versus Various Types

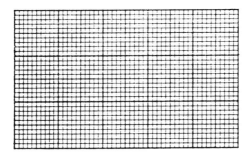

**Figure 2.** Real Values of Output Impedance versus Types (see text)

a. The action of the Lavisher may be smoothed somewhat by applying a drop or two of non-gumming KS-7415 Guile in the duct provided for that purpose.

**2.14** A typical response curve of a properly operating Regrettor is shown in Figure 3.

C. *Selectivity*

**2.15** In regard to selectivity, it may be said that the No. 2-B Regrettor affords a fine power of discrimination between regrets differing by as little as a single blue note. In fact, 5 moods off resonance the response is 60 db down.

D. *Output*

**2.16** During the recent lean years, many sluggards have repeatedly gone, not to the ant, but to the earlier types of regrettors which lacked sufficient output power to handle their requirements adequately.

**2.17** To fill this crying need, the No. 2-B Regrettor has been designed to handle a maximum output power of 60 watts R.M.S. (Room for More Surges). This is far in excess of the power normally required for ordinary household and industrial use.

a. It is of interest to note the fact that the No. 2-B Regrettor will withstand excessive overloads for long periods.

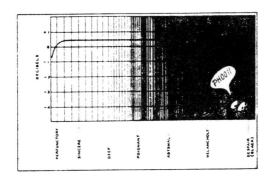

**Figure 3.** Typical Response Curve of No. 2-B Regrettor

Preliminary tests were made, in which the Republican Party Inc. and The House of Parliament Ltd. assisted in tests to destruction of the No. 2-B Regrettor. While regretting for the Republicans the test unit operated under conditions of almost 90% overload continuously for 6 months without failure, and this failure was directly traceable to an abnormally high concentration of hot gases and cigar smoke which interfered seriously with proper ventilation.

b. The results of the preliminary tests with the Parliament people, proved conclusively that the ultimate breakdown point of the instrument far exceeded the requirements of normal use. In this case the operating conditions called for a continuous headache punctuated at intervals by high amplitude pangs. The test was made during a time when there was some questions about the British coronary incumbent. The average overload was 72% and failure occurred by corona discharge and flash over in the plate circuits of the output tubes.

E. *Circuit Design Considerations*

**2.18** Spurious or parasitic regrets are prevented by the use of Ayrton-Perry windings on the beductance coils. The fact that each turn of wire is doubled back on itself is believed to be responsible in large measure for the low reactances of these coils and consequently for their effectiveness.

**2.19** The No. 2-B regrettor is always to be operated at the highest possible sobbing point. Sobbing point measurements can be made from readings of the non-rectifying type meter on the panel. This meter is of the past mistakes type having 1000 ohs per volt and will give a true indication of how far things have gone.

Note: Under conditions where regrets are desired in the 'acheing' range (4.5 to 5.6 moregrets/sec) a slight inherent tendency toward regeneration would set up a violent sobbing condition which might increase in

amplitude up to the peak capacity of the tubes were it not for the anti-sob resistors connected in the grid circuits. These effectively damp out all but the 'racking' type of sob.

a. The dampness is controlled by a lachrymal potentiometer conveniently located on the front panel.

**2.20** Since certain regrets call for a sustained racking sob condition, provision is made for dissipating the excess energy thus developed in a load circuit made up of four 600 ohm Utter Futility type pads.

**2.21** All grid prejudice voltages for the tubes are supplied from a single bias rectifier supply.

F. *Power Requirements*

**2.22** The No. 2-B Regrettor is designed to operate from a 105 to 120 volt 50 or 60 cycle A.C. or D.C. circuit and requires 100 watts.

G. *Mechanical Arrangement*

**2.23** The No. 2-B Regrettor unit is relatively light in weight, moderately small in size, fairly mobile and essentially the type of apparatus known as portable. A typical installation is pictured in figure 4.

**2.24** Overall dimensions are approximately as follows:

(a) Width: 24 33/64 inches
(b) Length: 60 63/64 inches
(c) Height: 40 37/64 inches

Note: These dimensions are overall and include the carrying handles.

Caution: *The handles are not furnished with the unit and must be ordered separately.*

(1) Bronze only is furnished when specified.

Note: Nickel screws only are furnished with handles.

(1) Screws should match the handles.

Note: Screws should be ordered separately.

**2.25** The total weight of the instrument including the No. 221-C Mounting and the No. D-9445 Anticipatory Converter Attachment is approximately 1242 pounds, 8 oz., 10 grams, 1 grain.

**Figure 4.** Showing Simplicity of Operation of No. 2-B Regrettor

**2.26** Controls are brought out to the front panel for accessibility. For ready maintenance on the interior of the cabinet, the 25 knobs and 15 keys are removed and the front panel is then removed by unscrewing 112 machine screws, and running an oxyhydrogen torch along the edges of the panel.

Caution: Since all wires and component units mounted inside the box are embedded in sealing compound, a trough should be placed below the front panel when it is removed.

### 3. Converter Attachment

**3.11** Due to the inherent principle of operation of the No. 2-B Regrettor, it assumes a perpetually pessimistic attitude. When the phase of this attitude is reversed, a steady state of optimism is set up. The D-9445 Anticipatory Converter Attachment is virtually a phase inverter. The pessimistic attitude of the Regrettor is reversed through a pi-section network.

**3.12** After reversal, the attitude is fed into a Class B attioamplifier which excites a Class C Extrapolator circuit. This extrapolator by means of its dynatronic characteristic of negative resistance, absorbs the output of the attioamplifier faster than it can be supplied, and the net result is that a projection into the future is obtained. This projection amounts to negative regression, which is, of course, the same as anticipation. Figure 5 illustrates a typical curve of the operation of a regrettor with the Anticipatory Converter attachment.

### 4. Installation and Operation

**4.11** The Regrettor should be placed in as bright and cheery a location as possible. Dark, dank corners should be avoided inasmuch as the instrument generally surrounds itself with an atmosphere of gloom and where this cannot be dissipated rapidly, it builds up to an undesirable concentration in the vicinity.

**4.12** The maximum ambient attitude should not be allowed to become higher than a medium 'pall of gloom'. Ambient attitudes higher than this produce a corrosive degenerative effect on the instrument causing its output to drop off, and ultimately the machine will go into a decline and waste away in the manner of an exponential function.

**4.13** The Regrettor should be mounted vertically.

a. When mounted horizontally the grid leak resistors cannot drain rapidly enough and the result is as shown in figure 6.

Note: See Figure 6.

(1) It is to be understood that this is representative of an average case only.

(a) Average cases are difficult to find.

Note: It is doubtful if such exist.

(1) See figure 6.

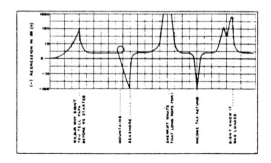

**Figure 5.** Regrettor Operation—Anticipator ON

**4.14** The regrettor should be turned 'ON'. Shortly the tubes will be hot, and the Regrettor will be prepared to receive its first assignment.

**4.15** The assignment generally used to determine that the set is functioning properly is essentially that of regret at having purchased the instrument. The procedure below would be followed:

a. Set switch D2 to the 'Regret' Position.

b. Advance 'OUTPUT' control to about half full scale.

c. Rotate the QUALITY dial to the setting, marked 'PANGS'.

d. Look at invoice reading 'Price: $3,145.25, slightly higher west of Kearney'.

e. Observe that the Regrettor has a single high amplitude pang, and note the immediate cessation of doubt and regret at having bought the instrument.

Note: (a) If it is found that following this pang, a wish to buy another identical unit arises, the OUTPUT control was set too high.

(b) If a slight lingering feeling of regret is noticed, the OUTPUT control was set too low.

**4.16** The D-9445 Anticipatory Converter should be attached and tested by patching through from jacks marked 'ANTIOUT' to 'REG. IN.' and from 'UT.FUT.MON' to 'ANT.HYB.IN' and the following procedure carried out:

a. Advance Regrettor OUTPUT to FULL.

b. Set QUALITY Knob to 'ANYTHG CN HPPN'.

c. Turn the OFF-ON switch on the Converter to 'ON' and observe that the Agastat S-3 picks up.

d. Rotate the Converter dial marked 'TIME' and its Vernier, 'HOURS', slowly, and observe the following future trends:

(1) Skirts will become longer, then much shorter.

(2) Non-fixable traffic tickets will be fixed.

(3) The New Deal will conduct investigations.

(4) Rows of theater seats will continue to be too close together.

**4.17** If it is observed that 'Hearst printspaper in RED ink', the instruments are defective and should be returned to the manufacturer for adjustment.

## 5. Calibration

**5.11** The calibration of the No. 2-B Regrettor is most conveniently checked by the arbitrary values of regret felt on March 15th. The desirability of the type of regret connected with this date lies in the fact that it is a universal standard, welling up on every hand, and flavored with equal bitterness everywhere. The following procedure is recommended in calibrating the No. 2-B Regrettor.

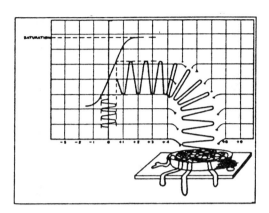

**Figure 6.**

a. Set the regrettor QUALITY dial to ACHEING and leave D-1 turned 'OFF'.

b. Look at notice from Bureau of Internal Revenue and recall 1936 Presidential Campaign.

c. Observe dull ache near billfold.

d. Suddenly turn Regrettor switch D-1 to ON and observe that as tubes heat and regrettor takes up the load, ache near billfold subsides, and a wish is felt to rise and sing first two verses of 'We're With You, Mr. President'.

**5.12** This procedure will accurately check the calibration. If the instrument is found to vary, repeat procedure and reset QUALITY dial to ACHEING.

## 6. THEORY

**6.11** From Durands' *Critique of Pure Reason,* it can be shown that, given sufficient time, an element $dv$ of the regression characteristic is a periodic function:

$$F_d(r) = FDR \sin \omega t + P^2W^2A^2(k).$$

The last term is largely transient as shown by its factor:

$$P^2W^2A^2 = (PWA)(WPA)$$

The summation

$$\sum_{'33}^{'37} (PWA)(WPA) = \frac{FDR + NRA}{8 \times 10^{-9}} \text{ dollars}$$

or

$$\lim_{t = 1940 \text{ AD}} f(P^2W^2A^2) = \text{G. O. P.} \times 10^8.$$

The latter half of this expression cancels out, of course, where the bias voltage of the Chief Justice approaches cutoff, and where the ratio 5/4 holds.

Since $\int_{MON}^{SAT} \phi \dfrac{g_0}{NRA}$

= a constant (Kerr–Plänck constant)

and

$$R = \pm j \frac{de}{dt} + \epsilon^{(d + j\beta)} - \mu L$$

it is clear that $\mu L$ in F. P. C. units (Farleys per cm$^2$) will be

$$\neq \frac{\sinh \propto + \cosh r + 2(a + j\beta)(.866 \lg)\log \epsilon^x}{A}.$$

The tippling coefficient as shown by Haig and Haig is

$$T = K(1 + .00073LV) \text{ per cavort}$$

which puts the quart before the hearse.

And the regrettor dynamic function is

$$T \times 10^8 + \frac{\partial R}{\partial T}.$$

This explains why $\int F(r)$ is essentially not linear.

Knowing $-\Delta R/\Delta T$, the characteristic can be extrapolated with a fair degree of accuracy to a first approximation.

Hence the gullibility factor ($\psi$) may increase without limit, and at all times equals the number of what $P^2T \times 10^9$ Barnum said was born every minute.

WHEEEE!! I'M HERMITIAN!!

# An archaeological hoax

Melin E Jahn and Daniel J Wolff 1965 *The Lying Stones of Dr Beringer* (Berkeley: University of California Press).

Pictured here are some fossil forms discovered in the 18th century by the German physician – geologist Johann Bartholomew Adam Beringer, a transitional figure in the history of geology. Unaware that these exotic 'fossils' had been placed in his dig by envious colleagues at the University of Würzburg, Beringer published a treatise, *Lithographiae Wircenburgensis*, to interpret them. His discovery of a 'fossil' bearing his own name made him aware of the hoax.

Beringer attempted to buy back all copies of his recently published book (1727), ruined himself financially, and shortly died of chagrin and mortification.

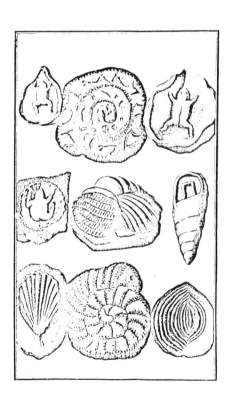
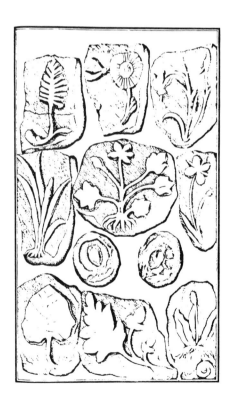

# Some hoaxes in the medical literature

RBP from Saul Jarcho 1959 *Bulletin of the History of Medicine* **33** 342–7. © The Johns Hopkins University Press. Excerpts, cf. *Some hoaxes in medical history and literature*.

[Address presented at the dinner session of the thirty-second annual meeting of the American Association for the History of Medicine, Cleveland, May 22, 1959.]

The arrival of a letter of invitation from the chairman of our Program Committee, Miss Elizabeth H Thomson, aroused doubts and raised difficulties. In search of guidance I turned to the King James Bible. The text fell open at Judges IV. 20. This part of Holy Writ presents the homicidal story of Sisera and Jael, the wife of Heber the Kenite. The line on which my finger rested reads as follows: '... and it shall be, when any man doth come and enquire of thee, and say, Is there any man here? that thou shalt say, No'. This seemed clear enough. But Miss Thomson's letter of invitation was so graciously worded that it could not be denied precedence over the Law and the Prophets.

It is an axiom of civilized life that the members and guests of a learned society who had just enjoyed a large dinner shall not have their post-prandial comfort troubled by a prolix oration, nor are they to be weighted down with metaphysical profundities. In compliance with this principle the address that you are about to hear will be brief.

As added evidence of the speaker's goodwill, there will be no mention of historiography, philosophy, sociology or economics, and nothing will be said about Florence Nightingale, Dorothea Lynde Dix, Ella Wheeler Wilcox, the ether controversy, antivivisectionism, or social insurance; and if Hippocrates is mentioned, this will be done in order to cast the mantle of antiquity and respectability upon the discussion.

The subject of this brief talk is, 'Some Hoaxes in the Medical Literature'.

It is not necessary at this time to seek the ultimate origins of the hoax. Anyone who reads Pliny must surmise that the credulous Roman had suffered imposture, and the same may have been true of Herodotus, if not Aelian.

In more recent periods the eighteenth century seems to have been especially productive of fraudulent and semi-fraudulent writings, as we know from Defoe's *Journal of the Plague Year*, Dean Swift's Bickerstaff papers, Macpherson's *Ossian*, and similar fabrications.

Coming to times still nearer to our own, we have the illustrious examples of William Henry Welch and Sir William Osler. Welch did not scruple to pretend that he suffered from aortic aneurysm, and on one occasion he went so far as to conceal a rubber bulb in his clothing in order to alarm Halsted [1].

In one recorded instance Welch proved himself, conversely, well able to resist imposture. The occasion was a meeting of the Association of American Physicians at which Abraham Jacobi [2] in complete good faith presented the case of a young man whose temperature allegedly had reached 148°. Jacobi moreover supported his statements by citing additional 'hot cases' from the literature. In the discussion Welch rejected Jacobi's report as incredible. In addition Welch told of another case of alleged hyperthermia—this one had been published in the *Journal of the American Medical Association* [3]—in which a temperature reading of 171°F had been obtained. These were cases of psychopathic patients reported by credulous physicians. But Welch, willing as he was to play practical jokes in private life, showed that he himself was proof against the tricks of hysterical malingerers.

My own interest in hoaxes arose in the following way. Many years ago I belonged to a journal club in a large eastern city. At one of our meetings a member reported that in a certain well-known clinical journal he had come across an article which was so profound as to be utterly unintelligible. When the text was produced for inspection, the leader of the club surmised that the article might be a hoax. Without intending to prejudice the case, I present a few excerpts herewith:

> . . . the methods assuring the utmost efficacy in prevention and treatment of antigenic and non-antigenic diseases are ultimately disclosed by hemology, which includes hemo-lympho-plasmology and hemo-lympho-cytology; i.e., the structure, functions and modes of production and storage of each non-cellular and cellular constituent of blood and lymph and knowledge of

activities responsible for wholesome and unwholesome modes of distribution of blood and lymph . . .

Antigenic influences, exerted from establishment of a fetal circulation until death, are the entrance into internal environment from tissue or from external environment (maternal circulation in fetus) or irritating cellular constituents and viruses (parasites) or non-cellular constituents (foreign proteins; i.e., parasitic and non-parasitic antigens) which are, despite filtration by vascular endothelium, lymph glands, lungs, liver, spleen, bone marrow and kidneys, rapidly disseminated in some measure throughout the body . . .

Whether or not this egregious text was *intended* to be unintelligible, it at least opened one's eyes to new vistas and new possibilities in the medical literature.

The subject of spurious medical writing came to attention again as recently as 1957 in the form of the Coudé hoax. For those who are not physicians I may be permitted to explain that a certain lack of straightforwardness in the structure of the human male has made it advantageous for urologists to employ an instrument which is slightly bent. From the French word *coude* (elbow), this bent device has come to be known as the *coudé* catheter.

In 1957 a journal named *The Leech*, which is published by the students' society of the Welsh National School of Medicine at Cardiff, printed the biography of one Dr Emile Coudé (1800 – 1870), alleged inventor of the *coudé* catheter [14]. This biography was equipped with footnotes and was ornamented by an impressive photograph—all spurious.

At this time Sir Hamilton Bailey was preparing a new edition of his *Short Practice of Surgery*. As he later reported in letters to the *Lancet* and the *British Medical Journal* [15], he did not detect the falsity of the *Coudé* biography until his text was in page proof. Consequently several pages of the new edition had to be reset and biographical references to Coudé deleted.

This incident caused a flurry of letters in the British medical press [16]. The *Lancet* carried the indignant complaint of one Hercule Coudé, who proclaimed that the honour of his illustrious granduncle Emile had been impeached [17].

*References*

[1] Flexner S and Flexner J T: *William Henry Welch and the Heroic Age of American Medicine* (New York: Viking Press) 1941, p 173.

[2] Jacobi A: Hyperthermy in man up to 148°F. (64.4°C.) *Trans. Assoc. Amer. Phys.*, 1895, **10** 159–189. Discussion by W H Welch and others, pp 189–191. *See also* Burket W C (ed) *Papers and Addresses by William Henry Welch* (Baltimore: Johns Hopkins Press) 1920, vol 1, pp 367–9.

[3] Galbraith W J: A remarkable case. *J.A.M.A.*, 1891, **16** 407–9. *See also* Summers J E Jr: Omaha's 'Remarkable Case' of high temperature. *Omaha Clinic*, 1891, **4** 115–8. *See also* Poe C T: *ibid.*, 1891, **4** 269–70.

[4] Davis E Y (pseudonym of William Osler): Vaginismus. *Medical News*, 1884, **45** 673. Not unexpectedly, this item is unlisted in Miss Minnie Blogg's *Bibliography of the Writings of Sir William Osler*. Revised edition. Baltimore, 1921, 96 pp. *See also* An uncommon form of vaginismus (an editorial), *Medical News*, 1884, **45** 602–3.

[5] Champneys F H: Notes of a clinical lecture on vaginismus. *Clinical Journal*, 1892–3, **1** 42–4.

[6] Wilson J G and Fiske J: *Appleton's Cyclopaedia of American Biography* (New York: Appleton) 1887–1900. 7 vols.

[7] Winchell C M: *Guide to Reference Books*. Seventh edition. (Chicago: American Library Association) 1951, p 434. *See also* '84 Phonies.' *Letters*, 1936, 3(19) 1–2. Schindler M C: *Fictitious biography. Am. Hist. Rev.*, 1937, **42** 680–90. MacDougall C D: *Hoaxes*. Second Edition. (New York: Dover Publications) 1958, p 227.

[8] Hirsch A: *Handbook of Geographical and Historical Pathology*. (London: New Sydenham Society) 1882–1886, vol 1, pp 401, 421.

[9] Originally published under the title of 'A neglected anniversary' in the *New York Evening Mail* of December 28, 1917. *See* Mc Hugh R (ed): *The bathtub hoax . . . by H. L. Mencken*. (New York: Knopf) 1958.

[10] MacDougall C D, *op. cit.*, pp 304–9.

[11] Zinsser H: *Rats, Lice and History*. (Boston: Little, Brown) 1935, p 285.

[12] Young H: *A Surgeon's Autobiography*. (New York: Harcourt, Brace) 1940, pp 64–5.

[13] Heiser V: *You're the Doctor*. (New York: Norton) 1939, p 228.

[14] R P: Emile Coudé (1800–1870), *The Leech*, 1957, 4(6) 15–6.

[15] Bailey H, Bishop W J and Morson A C: Coudé straightened out. *Lancet*, 1958, **2** 424, and *Brit. M. J.*, 1958, **2** 513.

[16] See *Lancet*, 1958, **2** 522, 526; *Brit. M. J.*, 1958, **2** 567, 640, 693. My thanks are offered to Mrs. Elizabeth Bready of the New York Academy of Medicine for these entertaining items.

[17] *Lancet*, 1958, **2** 456.

[18] Gano S N: A gloss attributed to the Hippocratic school. *Leech*, 1958, 4(8) 17–9.

# The moon hoax

RBP from S A Mitchell's article in *Popular Astronomy* summarized in *The Observatory* **23** 260 (1900).

In August of the year 1835 a series of articles appeared in the New York *Sun* purporting to be a description of Sir John Herschel's observations at the Cape: it will be remembered that Sir John landed in South Africa in January 1834. These articles begin in a *quasi* scientific manner, describing plans said to have been invented by the elder Herschel for constructing improved reflecting telescopes, which failed only because the size of the image was such that it was not sufficiently illuminated, and then go on to state that Sir John Herschel had evolved the idea of *lighting the focal image by artificial light* so that the faintest image formed in the focus of a telescope could be seen. Description was then given of an enormous instrument made by Sir John on these lines, and of the wonderful things he saw on the Moon by its aid—magnificent animals, creatures like men with wings, and so on. This fable appears to have met with considerable success; for one thing it considerably increased the circulation of the journal in which it appeared, and also it is said to have received some credence from Arago, who circulated the wonderful account through Paris. This last fact gives Mr Mitchell a clue to the real authorship. He thinks that the actual originator of the 'Moon Hoax' was Nicollet, an astronomer, who had left Paris in some disrepute, and did this, first to earn some money, and, secondly, to entrap his enemy Arago into a foolish position.

\*\*\*\*\*\*

From *The Observatory* **23** 295 (1990)

... Mr S A Saunder writes to us with reference to ... last month's number in which it appears that we did considerable injustice to Arago. Mr Saunder ... says that when the popular excitement was at its height, Arago, who was then President, read extracts from the recently published 'Moon Hoax' before the Academy of Sciences, and said that he thought the least that men of science in Europe could do was to prevent Sir John Herschel's name from being associated with such nonsense. He took the matter very seriously, and with tears in his eyes besought the Academy to formally express its opinion to that effect, the result being that amid roars of laughter the account was formally declared to be incredible.

# An introduction to semiconductors

*J T Wallmark*
Research Laboratory of Electron Physics III,
Chalmers University,
Gothenburg, Sweden 41296

**A** is for ACCEPTORS, electrons like them best.
While the holes do the work the electrons rest.

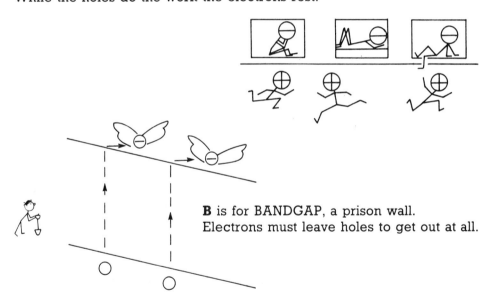

**B** is for BANDGAP, a prison wall.
Electrons must leave holes to get out at all.

**C** is for CHARGES, they are considered singularities,
but are very attractive when of different polarities.

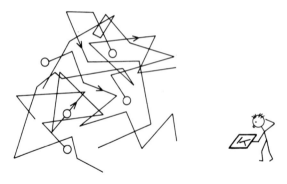

**D** In semiconductors, **D** is for DIFFUSION —
to elbow one's way through a crowd in confusion.

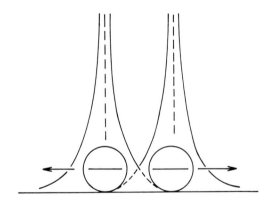

**E** is for ELECTRONS, their manner is quite coarse,
they shove each other with a repulsive force.

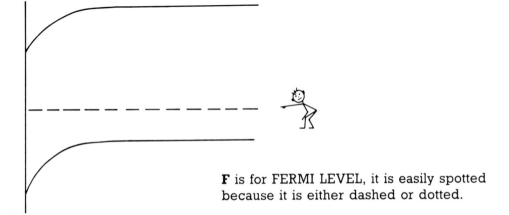

**F** is for FERMI LEVEL, it is easily spotted
because it is either dashed or dotted.

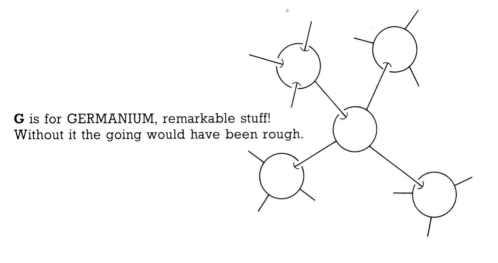

**G** is for GERMANIUM, remarkable stuff!
Without it the going would have been rough.

**H** A HOLE in a somewhat urban phrase
is an empty electron parking space.

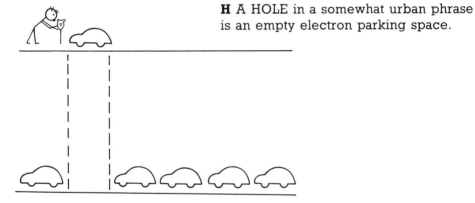

**I** is for INVERSION, a distinctly crooked scheme,
the minority takes over and rules supreme.

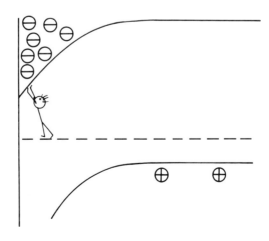

**J** is for JAPAN which is quite a solid state.
Electrons get along in Japanese just great.

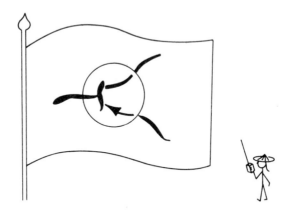

**K** is for *k*-SPACE, though too abstract for a swim,
with perfect waves it is filled to the brim.

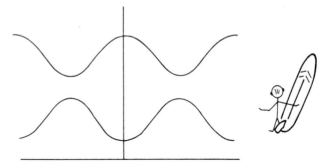

**L** is for LATTICE, described by a vector-set,
but actually looks as if built by an erector-set.

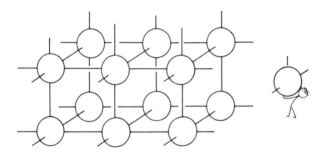

**M** According to MURPHY'S LAWS all efforts come to nil.
If something can possibly go wrong, it will!

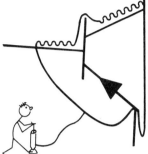

**N** In an N-TYPE medium you don't have much fun, each character you meet is a negative one.

**O** is for OXIDE, it grows very thin,
just enough for all surface effects to get in.

**P** PHOTONS are pretty — red, green or blue,
but PHONONS are more interesting to listen to.

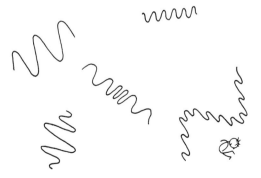

**Q** QUASIFERMILEVEL, — what a frightful word!
but considering the alternatives, it is to be preferred.

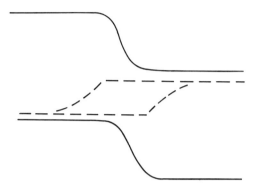

**R** If you look at a RECTIFIER at all philosophically,
it is where carriers disappear catastrophically.

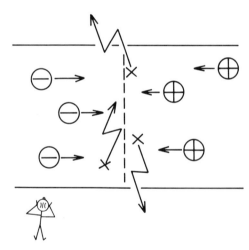

**S** is for SURFACE STATES, an excuse we get exposed to
when experiments won't work the way they are supposed to.

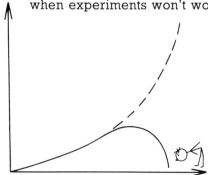

**T** is for TRANSISTOR, its success should not amaze,
it always had a very solid base.

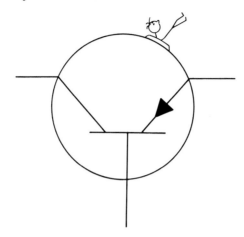

**U** UNIPOLAR means electrons aren't allowed when there are holes,
or, vice versa, with a change of signs, reverse the roles.

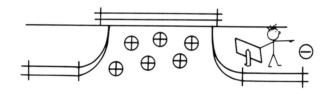

**V** is for VACANCIES. People insist
on discussing things that do not exist.

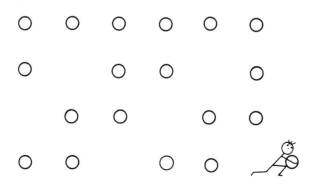

**W** This book could be the WALLMARK of civilization were it not for the whimsical versification.

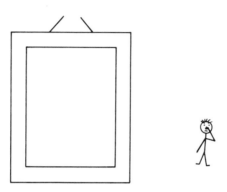

**X** is for X-RAYS, they are rather hard to scatter.
That's why they are useful to find out what's the matter.

**Z** is for ZENER DIODE — and now we are through with it —
except to figure out what Zener had to do with it.

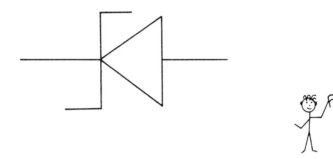

**Y** Production engineers have acquired immunity
from the law that YIELD and shrinkage must add to unity.

$$\begin{array}{r} .01 \\ + .36 \\ \hline 1.00 \end{array}$$

# Food and Environmental Concerns

Science is one thing, wisdom is another. Science is an edged tool, with which men play like children, and cut their own fingers. If you look at the results which science has brought in its train, you will find them to consist almost wholly in elements of mischief. See how much belongs to the word Explosion alone, of which the ancients knew nothing.

*Arthur Stanley Eddington*
*The Decline of Determinism*

# A ballad of energy policy

*Kenneth E Boulding*

RBP of K E Boulding. Presented at the WATTec Symposium on Energy and the Public, Knoxville, TN, 21–23 February 1979.

*Who from Ben Franklin has not learned,*
*A penny saved is a penny earned;*
*So it should be clear, when oil is short,*
*A barrel saved is a barrel bought.*
*And if we want to save the dollar,*
*'Conserve' is what we need to holler;*
*Though economic man won't keep,*
*Conserving things that are too cheap.*

*The answer is extremely clear,*
*That what is scarce should then be dear;*
*And if by subsidies we keep,*
*Things that are scarce, absurdly cheap;*
*Demand will bloat, supply will shrink,*
*Sooner than most of us might think;*
*So subsidies should all be axed,*
*And Btu's should all be taxed.*

*Though, lest we think the answer's easy,*
*All the simple answers make us queasy;*
*When prices equal social cost,*
*The bosses gain more than the bossed;*
*So policies are all too sure,*
*To bless the rich and screw the poor;*
*Because our system sadly lacks,*
*A real progressive income tax.*

*Prince, I don't envy you your task,*
*Questions that are not hard to ask;*
*Are very hard to answer—so,*
*The word's 'go carefully'—but go!*

404

# A ballad of ecological awareness

*Kenneth E Boulding*

RBP of K E
Boulding 1972
*The Careless
Technology:
Ecology and Inter-
national Develop-
ment* ed M Taghi
Farvar and John
P Milton (Garden
City, NY: The
Natural History
Press).

*Ecological awareness leads to questioning of goals:*
*This threatens the performance of some old established roles.*
*So to raise the human species from the level of subsistence*
*We have to overcome Covert Political Resistance.*
*So we should be propagating, without shadow of apology,*
*A Scientific Discipline of Poleconecology.*

*Among the very saddest of developmental tales*
*Is the indestructibility of fluke-infested snails.*
*Development is fluky when with flukes the blood is crammed,*
*So the more we dam the rivers, then the sooner we are damned.*

*Schistosomiasis has conquered—for the sad Egyptian fails*
*In six thousand years of history to eliminate the snails;*
*Yet in spite of all the furor of ecologist's conniptions*
*The Snail has failed completely to eradicate Egyptians.*

*In use upon the water of a good molluscicide*
*We really don't know what is true, but only what is tried.*
*For snails are pretty clever and climb upon the bank;*
*So if any good is done at all, we don't know what to thank.*

*Development must be successful, O, my darling daughter,*
*So keep your clothes on all the time, and don't do near the water.*
*The best advice we have is—for developmental tactics,*
*Don't wash or swim or go to bed without your prophylactics.*

*Bacteria have learned the trick of formal education—*
*They can transfer drug resistance with a shot of information.*
*So perhaps our universities should go in new directions,*
*And give their education by a series of injections.*

*The more we move around the world to where the prospect
pleases,*
*The more we will communicate deplorable diseases.*
*Yet there may be a solution if we do not choose to flout it,*
*If we can communicate just what to do about it.*

405

## A ballad of ecological awareness

Development will conquer the diseases of the poor,
By spraying all the houses and by putting in the sewer.
And we'll know we have success in our developmental pitch,
When everybody dies from the diseases of the rich.

The cost of building dams is always underestimated—
There's erosion of the delta that the river has created,
There's fertile soil below the dam that's likely to be looted,
And the tangled mat of forest that has got to be uprooted.

There's the breaking up of cultures with old haunts and habits
    loss,
There's the education program that just doesn't come across,
And the wasted fruits of progress that are seldom much enjoyed
By expelled subsistence farmers who are urban unemployed.

There's disappointing yield of fish, beyond the first explosion;
There's silting up, and drawing down, and watershed erosion.
Above the dam the water's lost by sheer evaporation;
Below, the river scours, and suffers dangerous alteration.

For engineers, however good, are likely to be guilty
Of quietly forgetting that a river can be silty,
While the irrigation people too are frequently forgetting
That water poured upon the land is likely to be wetting.

Then the water in the lake, and what the lake releases,
Is crawling with infected snails and water-borne diseases.
There's a hideous locust breeding ground when water level's
    low,
And a million ecologic facts we really do not know.

There are benefits, of course, which may be countable, but which
Have a tendency to fall into the pockets of the rich,
While the costs are apt to fall upon the shoulders of the poor.
So cost-benefit analysis is nearly always sure,
To justify the building of a solid concrete fact,
While the Ecologic Truth is left behind in the Abstract.

By undiscriminating use of strong insecticide
Our temporary gain is lost when all our friends have died.
With strip planting of alfalfa something new is making sense:
Spend the millions now on tribute—not a penny for defense!

The locust as an insect is extremely international,
It runs a downwind airline, it's adaptable and rational.
For biological controls the beast is far too mobile;
It seems a shame to persecute an animal so noble.
But though the locust is a most engaging little rascal,
I think I'd put my money on Ecology and Haskell.

One principle that is an ecological upsetter
Is that if anything is good, then more of it is better,
And this misunderstanding sets us very, very wrong,
For no relation in the world is linear for long.

Pursuit of agriculture on a lateritic soil
Is a classical example of an Unrewarding Toil,
For the unsuspecting settler gets a very nasty shock
When the lateritic soil turns into lateritic rock.

The poisoned mouse eliminates the useful owl and vulture,
But the growing world economy insists on monoculture.
O! Science may be phony but the social system's phonier,
And so spread on, insecticide, and sulphate of ammonia.

A developed Agriculture is a fabulous polluter;
As development gets faster, then the problem gets acuter.
We are loading up the planet with a lot of nitric trash,
And if nitrogen falls off its cycle—wow! is that a crash.

Development is fatal to the local and specific;
A single culture spreads from the Atlantic to Pacific.
So preserving every specimen of life is quite essential
If we're not to break the bank of evolutionary potential.

Too many governments, alas, in tropic parts today
Say, 'Let us group the little farms', and then say, 'Let us spray'.
Pests and Pollutions prosper then, and what is more the pity,
It drives the people off the land to fester in the city.
So how do we inculcate in a heterogeneous nation
The sober sense of ignorance that leads to conservation?

A cost of exercising power is much unwanted heat,
So victory is found to be a species of defeat,
And no amount of slick brochures can cultivate a taste
For dead and tepid rivers and for radioactive waste.

*The growth of population has a great deal of momentum,*
*Neither spirals, interruptus, or safer still, absentum*
*Can do much about the kids who are already on the scene,*
*Who will still be in the labor force in twenty-seventeen.*
*So there isn't very much that the developed world can do*
*To help the poor old woman in the very crowded shoe.*

*The oceans we have mobilized to feed the too-well-fed;*
*The rain is red in Adelaide from deserts newly bred;*
*We nibble at the nomads, though oil-rich and water-poor;*
*Displaced Masai, domesticated, have a drought in store.*

*The tsetse fly can guard the wild, as long as it survives;*
*As men and cattle press the land, the game no longer thrives;*
*The tourist business is a trap, it is a tainted honey;*
*Man clearly should have stayed in bed, and not invented money.*

*It's nice to be the drafter of a well-constructed plan,*
*For spending lots of money for the betterment of Man,*
*But Audits are a threat, for it is neither games nor fun*
*To look at plans of yesteryear and ask, 'What have we done?'*
*And learning is unpleasant when we have to do it fast,*
*So it's pleasanter to contemplate the future than the past.*

*If it's just the noise of progress that is beating in our ears*
*We could look beyond the turbulence and sooth our gnawing*
   *fears.*
*Man is drowning in his own success, and hapless is his hope*
*If our science and technology is but a rotten rope.*

*Infinity is ended, and mankind is in a box;*
*The era of expanding man is running out of rocks;*
*A self-sustaining Spaceship Earth is shortly in the offing*
*And man must be its crew—or else the box will be his coffin!*

408

# Conservation versus technology

K E Boulding 1956 in *Man's Role in Changing the Face of the Earth* ed W L Thomas (Chicago: University of Chicago Press) p 1087.

## A Conservationist's Lament

*The world is finite, resources are scarce,*
*Things are bad and will be worse.*
*Coal is burned and gas exploded,*
*Forests cut and soils eroded.*
*Wells are dry and air's polluted,*
*Dust is blowing, trees uprooted.*
*Oil is going, ores depleted,*
*Drains receive what is excreted.*
*Land is sinking, seas are rising,*
*Man is far too enterprising.*
*Fire will rage with Man to fan it,*
*Soon we'll have a plundered planet.*
*People breed like fertile rabbits,*
*People have disgusting habits.*

**Moral:**

> *The evolutionary plan*
> *Went astray by evolving Man.*

## The Technologist's Reply

*Man's potential is quite terrific,*
*You can't go back to the Neolithic.*
*The cream is there for us to skim it,*
*Knowledge is power, and the sky's the limit.*
*Every mouth has hands to feed it,*
*Food is found when people need it.*
*All we need is found in granite*
*Once we have the men to plan it.*
*Yeast and algae give us meat,*
*Soil is almost obsolete.*
*Men can grow to pastures greener*
*Till all the earth is Pasadena.*

**Moral:**

> *Man's a nuisance, Man's a crackpot,*
> *But only Man can hit the jackpot.*

# God and EPA

Andrew J Hinshaw *Congressional Record* October 10, 1974 and *CHEMTECH* 8 October 1978 IBC. Published 1978 by the American Chemical Society.

In the beginning God created Heaven and Earth.

He was then faced with a class action lawsuit for failing to file an Environmental Impact Statement with HEPA (Heavenly Environmental Protection Agency), an angelically staffed agency dedicated to keeping the Universe pollution free.

God was granted a temporary permit for the heavenly portion of the project but was issued a Cease and Desist Order on the earthly part, pending further investigation by HEPA.

Upon completion of his Construction Permit Application and Environmental Impact Statement, God appeared before the HEPA Council to answer questions.

When asked why he began these projects in the first place, he simply replied that he liked to be creative.

This was not considered adequate reasoning and he would be required to substantiate this further.

HEPA was unable to see any practical use for earth since 'The earth was void and empty and darkness was upon the face of the deep'.

Then God said: 'Let there be light'.

He should never have brought up this point since one member of the Council was active in the Sierrangel Club and immediately protested, asking 'How was the light to be made? Would there be strip mining? What about thermal pollution? Air pollution?' God explained the light would come from a huge ball of fire.

Nobody on the Council really understood this, but it was provisionally accepted assuming (1) there would be no smog or smoke resulting from the ball of fire, (2) a separate burning permit would be required, and (3) since continuous light would be a waste of energy it should be dark at least one-half of the time.

So God agreed to divide light and darkness and he would call the light Day, and the darkness Night. (The Council expressed no interest with inhouse semantics.)

When asked how the earth would be covered, God said, 'Let there be firmament made amidst the waters; and let it divide the waters from the waters'.

The ecologically radical Council Member accused him of double talk, but the Council tabled action since God would be required first to file for a permit from the ABLM (Angelic Bureau of Land Management) and further would be required to obtain water permits from appropriate agencies involved.

The Council asked if there would be only water and firmament and God said, 'Let the earth bring forth the green herb, and such as may see, and the fruit tree yielding fruit after its kind, which may have seen itself upon the earth'.

The Council agreed, as long as native seed would be used.

About future development God also said: 'Let the waters bring forth the creeping creature having life, and the fowl that may fly over the earth'.

Here again, the Council took no formal action since this would require approval of the Game and Fish Commission coordinated with the Heavenly Wildlife Federation and Audobongelic Society.

It appeared everything was in order until God stated he wanted to complete the project in 6 days.

At this time he was advised by the Council that his timing was completely out of the question ... HEPA would require a minimum of 180 days to review the application and environmental impact statement, then there would be the public hearings.

It would take 10 to 12 months before a permit could be granted.

God said, 'To Hell with it!'

# Tuna on rye, 1984

*Samuel Vaisrub*

RBP from *JAMA* **243** No 18, May 9, 1980, p 1846. Copyright 1980, American Medical Association.

'Your medical clearance, sir?'

'My what?'

'Your clearance, of course', repeated the waiter impatiently. 'Didn't you order an omelette and buttered toast?'

'That I did, but what has it to do with a clearance?'

The look of incredulity on the waiter's face gradually gave way to that of understanding. 'You must be a stranger in these parts.'

'That I am. I am also hungry. What gives with the clearance bit?'

'All we need is the record of your serum low-density lipoproteins. You can't have fatty food otherwise.'

'How about a corned beef sandwich', I asked timidly. 'No butter.'

'Sorry, there is a lot of saturated fat in corned beef.'

'Smoked salmon, perhaps?'

'Out of the question. You must have clearance for high blood pressure before we can serve you salty food.'

'Could you just give me a piece of apple pie and some cheese', I asked with what must have sounded like the voice of despair. 'After all, you remember what Eugene Field said:

> No matter what condition
> Dyspeptic come to tease
> The best of all physicians
> Is apple pie and cheese.'

'No', persisted the adamant waiter, 'not until you provide documentary proof that you don't have diabetes.'

At this point I was desperate, and I began to think in terms of crime—violence? theft? bribery? 'What would happen', I asked, rolling a $10 bill between my fingers, 'if I slipped you this piece of greenery?'

'I would immediately report you to the Gastronomic Department of the CIA. To bribe a waiter is a more serious offense than to tip a police officer.'

'Would you please hand me the menu?' I asked, playing for time. Somehow, I had to get a bite of food into my empty stomach.

The menu was huge, containing pages of à la carte items with detailed breakdown of their mineral and vitamin content, as well as their caloric value and medical clearance requirements. It was impossible to wade through the culinary morass. As I was about to give up hope, I heard a whisper.

'Can I interest you in a tuna on rye?' The voice was that of the occupant of the adjacent table. 'I'll give it to you, if you slip me the ten-spot that you're fondling with your right hand.'

The transaction was completed in a matter of seconds and the sandwich devoured in a few minutes. Speed was essential, if I was to avoid a confrontation with the returning waiter.

'Did you decide on your order, sir?' asked the waiter seemingly more sympathetic to my predicament.

'No. I can't find anything that does not require a medical clearance.'

'I wouldn't say that', beamed the waiter. 'We always feature one clearance-free dish, which is both cheap and harmless. Today it is a 50-cent item—tuna fish on rye.'

## Obsolescence

RBP of Frances Weinberg.

*How simple life once used to be,*
*So nonchalant, and so carefree.*
*We dwelt on the complacent side*
*Then heard about thalidomide.*

*Of many hazards science spoke:*
*Carcinogens and cigarette smoke,*
*Grave warnings about DDT,*
*And tuna fish with mercury.*

*While wondering what else was at stake*
*We parted with sweet cyclamate.*
*The phosphate perils they came next,*
*Now saccharin has failed the test.*

*Of each new banning that we learn,*
*There is one of utmost concern,*
*How will the human race stay clean...*
*Without our hexachlorophene.*

## Cartoons and the environment

[Witty World is primarily an international trade magazine for the cartoon industry. It has a staff of 56 living in 54 countries and claims readers in 103 countries. Many articles have treated the comic profession in various countries and have touched on issues such as censorship, 'cheapening of the comics', literacy and comics, graffiti, and use of comic art for social change. Here is a sample of an article on air and water polution, waste disposal, and deforestation.]

414

RBP from Brian
D Harding. ©
1987 WittyWorld
International Car-
toon Magazine,
No 2, Autumn
1987 p 18.

**FANG CHENG/
People's Republic
of China**

*"These are what I get from this river since your factory
started production."*

**JAROSLAV DOSTAL/Czechoslovakia**

WittyWorld International Cartoon Magazine, No. 2, Autumn 1987

415

**LÁSZLÓ DLUHOPOLSZKY/Hungary**

WittyWorld International Cartoon Magazine, No. 2, Autumn 1987

# Looking Back

The wife of the Bishop of Worcester after the
Huxley – Wilberforce debate: 'Descended from the
apes! Let us pray it is not true, and if it is, let us pray
that it does not become generally known.'

M Ruse

# On fashions in science

*Sir Ernst Chain*

RBP from
*Biotechnological
Applications of
Proteins and
Enzymes* ed Zvi
Bohak and
Nathan Sharon
(New York:
Academic Press)
1977, p 15.
Copyright © 1977
by Academic
Press, Inc.

Some trends of non-applied research are dominated by fashion, and, strangely enough, these fashions have as strong an influence on the scientist as they do on a woman in the choice of a dress. Work on the biochemical basis of memory is a good example of this. It has swept neurobiological laboratories like a viral epidemic, has swallowed up large funds for absurd experiments, and has produced few, if any, results of any interest. Fortunately, this epidemic seems to have abated. Similar fashions can be found in some areas of the field known under the misnomer of molecular biology. A great deal of work on biological membranes and on molecular genetics, particularly genetic engineering, belongs in this category. In the former, a mass of data is obtained by the application of new physical measuring techniques, such as nuclear magnetic or electron spin reasonance, or by studying the amino acid composition of some proteins or the structure of lipids in some membranes of some cell types. The physiological meaning of such data cannot be interpreted, nor is there any likelihood that a reasonable interpretation will be possible in the foreseeable future. Papers in this field fill, however, several periodicals specifically devoted to the publication of membrane studies; many of them read like extracts from rather poor PhD theses.

Predictions that human genes coding for peptide hormones, such as insulin, can be obtained from the chromosomes by specific excision or by synthesis, can be transferred into bacterial genomes, and will then direct their protein synthesis specifically toward the production of the peptide hormones so that they can be obtained in large amounts by fermentation techniques seem to me to belong at the present time more in the field of science fiction than science.

418

# Remarks on quarks

*Reinhold Gerharz*

RBP from
*American Scientist,*
Journal of Sigma
Xi, The Scientific
Research Society,
Winter 1968,
p 425A.

In an interesting sequence of letters, Epstein, Wilson, and Weichert [1] explored and uncovered the traces in literature which lead through James Joyce's *Finnegans Wake* to the true identity of that abominable 'quark'-man: it is the German writer and scientist Johann W Goethe who, in his play *Faust* portrays a physicist [2] and reflects in him mankind's ever unsatisfied thirst for truth through carefully interluded pleasure.

These letters and Goethe's play are symbolic of the widening circle of confusion which impairs and intertwines the competences of Man and the providence of the Good Lord. Following the play through the dialogue between the Lord and chief devil Mephistopheles [3] it is the latter who notices Man behaving beastlier than beasts before ending his laments in great disgust with the pointed passage: 'In jeden Quark begrabt er seine Nase'. He then proceeds to capitalize on Man's fallibility which becomes substantiated in the plan by Man, prostituting his own soul, in the murder of his own kind, and his rape of nature.

As the philosophical images of all the characters are refocused in the opening and closing scenes of the play, Goethe (who otherwise never uses the same wording twice) makes the devil once more exclaim his malice after trying to catch the likeness of an angel: 'Du bleibst ... ein ekler Gallerquark!' [4].

We notice that the term 'quark' has found a celebrated rehabilitation after it became introduced into Elementary Particle Physics as Gell-Mann's Quarks [5]. In compliance with its meaning, the word has turned out to be an agent of tainture, fermentation, and spoilage. In many scientific papers it signals a peculiar decay in reasoning while it simultaneously creates bubbles of ignorance and depreciation. One begins to wonder whether Physics deserves such an injection of impurity.

Perhaps it would be wise to let this word remain within the exclusive repertoire of German cheese manufacturers, who have used this term long before Goethe borrowed it around 1780 for his parable.

*References*

[1] *Scientific American* September 1968 p 20
[2] W Heisenberg's interpretation in *Physikalischer Blätter* **24** Heft 4/5 p 196 (1968)
[3] 'Faust 1', verse 292
[4] 'Faust 2', verse 11742
[5] M Gell-Mann *Physics Letters* **8** p 214 (1964).

# Earth, heavens, and beyond

RBP from Edward Grant 1971 *Physical Science in the Middle Ages* pp 76–8. © Cambridge University Press.

Already in antiquity, certain Stoics, though in agreement with Aristotle that (1) the cosmos itself is a finite sphere without vacua (it was filled with pneuma) and that (2) all existent reality and matter were contained within it, insisted that our finite world was surrounded by an actually existent three-dimensional void capable of receiving matter and serving as its receptacle. The infinite void serves as a receptacle for our infinite cosmos. Indeed this seems to be its only function, since interaction between cosmos and infinite void is denied on grounds that the latter has no properties of its own and can in no way affect the material world, which is closed off and sealed from it. The world cannot be dissipated into the void. But granting that the world is surrounded by void, why, it was asked, must it be assumed infinite? In response, it was argued that because no body could exist beyond the physical world, no material substance could limit the void; and since it was absurd to suppose that void could limit void or that void should terminate at one point rather than another—a clear violation of the principle of sufficient reason—the infinity of void space seems an irresistible conclusion.

Although Stoic cosmology was hardly known in the Middle Ages, one significant argument was made available by Simplicius. In Moerbeke's 1271 Latin translation of Simplicius's commentary on Aristotle's *On the Heavens*, the Stoics are reported to have proved the existence of a vacuum beyond the world. What would happen they asked, if someone at the outermost extremity of the world extended his arm? Either the arm would reach outside the world from which it could be inferred that a vacuum lay beyond, or the arm would meet an obstacle in the form of matter, in which event the person must then stand at the extremity of the obstacle and again extend his arm. Since the world is assumed finite, this act can be repeated only a finite number of times. Eventually the arm will meet no obstacles at which point a vacuum may be inferred. Although this argument was cited by Aquinas, Buridan, Oresme, and others, its significance was overshadowed by a concept derived from another ancient source.

In an anonymous Latin Hermetic dialogue called *Asclepius*, written in the second or third century A.D., Hermes Trismegistus ('Thrice great Hermes') explains to Asclepius that if void existed beyond the cosmos, which he doubts, it would be void of physical bodies only but never of spiritual substances intelligible to the mind alone. Since *Asclepius* was known in the Middle Ages and Renaissance, it may have transmitted the concept of an extracosmic space filled with spirit but empty of matter, a concept destined to play a vital role in discussions about what might lie beyond the cosmos.

Its influence is detectable on Thomas Bradwardine, who, in his lengthy theological treatise, *De causa Dei contra Pelagium*, a defense of God against the Pelagians, furnished it with a Christian rationale based on God's infinite power and omnipresence. Assuming that God's perfection would be more complete if He existed in many places simultaneously than in a unique place only, Bradwardine demonstrates that God necessarily exists in every part of the world and also everywhere beyond the real world in an imaginary infinite void. Since God is omnipresent in an infinite void beyond the world, it follows—and here we detect the influence of *Asclepius*—

that although void can exist without body, it cannot exist without God's presence.

Theological considerations seem to have impelled Bradwardine to associate these empty places with God. Prior to the creation of the world, a place for it must have existed—indeed an infinity of possible places, since God could have created the world anywhere He pleased. If God had created the place, or possible places, of the world before He created the world itself, the former, not the latter, would have been the first creation and the uniqueness of the creation would be lost. To escape this dilemma, Bradwardine assumed that the infinity of potential places of the world is eternal and uncreated. Co-eternality with God, however, was not to be construed as implying independence from God. Such an interpretation would have signified the existence of two uncreated entities, and greatly diminished God's unique status. In some sense, then, these places and spaces must be associated, and perhaps even identified, with God. How this is to be understood was left unexplained. Are the places in God, or God in the places? Are the places attributes of God? Ignoring these and other questions, Bradwardine sought to avoid more obvious and immediate theological pitfalls while maintaining his conviction that God is omnipresent in an imaginary infinite place void of everything but the deity.

Despite the infinite omnipresence of God, Bradwardine cautions against describing Him as an 'infinite magnitude' in the ordinary sense of an extended, dimensional entity. God is infinitely extended only in a metaphysical sense without actual extension and dimension. Incomprehensible as this may seem, it is obvious that, to Aristotle's denial of extracosmic void, Bradwardine formulated a response based on contemporary Christian theology. In so doing, he placed himself in disagreement with such Church luminaries as Augustine and Aquinas, who had supported Aristotle's position. His conclusions were also at variance with the opinions of Duns Scotus (ca 1265–1308) and his followers, who agreed that God's will, not his omnipresence, was the basis of divine action. God could act on, and in, a place remote from his actual presence. It was, therefore, unnecessary to assume God's prior presence in the empty place where He created the world, for which reason Scotus denied the necessity of God's omnipresence in an infinite void space. God's presence in an infinite void beyond the world, as Bradwardine described it, was a new and significant element in medieval cosmology, one that stimulated further discussion in the late Middle Ages and Renaissance.

---

## Biology in verse

*In the beginning†*

Reproduced from Ralph A Lewin 1981 *The Biology of Algae and Other Verses.*

*In the beginning the earth was all wet;*
*We hadn't got life—or ecology—yet.*
*There were lava and rocks—quite a lot of them both—*
*And oceans of nutrient Oparin broth.*
*But then there arose, at the edge of the sea,*
*Where sugars and organic acids were free,*
*A sort of a blob in a kind of a coat—*
*The earliest protero-prokaryote.*
*It grew and divided: it flourished and fed;*
*From puddle to puddle it rapidly spread*
*Until it depleted the ocean's store*
*And nary an acid was found any more.*

*Now, if one considered that terrible trend,*
*One might have predicted that that was the end—*
*But no! In some sunny wee lochan or slough*
*Appeared a new creature—we cannot say how.*
*By some strange transition that nobody knows,*
*A photosynthetical alga arose.*
*It grew and it flourished where nothing had been*
*Till much of the land was a blue shade of green*

†R. A. L. 1977 *Biologist (J. Inst. Biol.)* **24** 10.

*And bubbles of oxygen started to rise*
*Throughout the world's oceans, and filled up the skies;*
*While, off in the antediluvian mists,*
*Arose a few species with heterocysts*
*Which, by a procedure which no-one can tell,*
*Fixed gaseous nitrogen into the cell.*

*As the gases turned on and the gases turned off,*
*There emerged a respiring young heterotroph.*
*It grew in its turn, and it lived and it throve,*
*Creating fine structure, genetics, and love,*
*And, using its enzymes and oxygen-2,*
*Produced such fine creatures as* coli *and you.*

*This, then, is the story of life's evolution*
*From Oparin broth to the final solution.*
*So, prokaryologists, dinna forget:*
*We've come a long way since the world was all wet.*

*We owe a great deal—you can see from these notes—*
*To photosynthetical prokaryotes.*

# Science and the citizen

## Voyager II

*Robert J Nash*

[CHEMunications is a publication of the Rochester Section, Inc, of The American Chemical Society. It has been traditional for one issue a year to be the 'humor' issue. During his tenure as editor, Robert J. Nash decided to parody several publications—*Scientific American*, the local Rochester newspaper, *National Geographic*, and *Playboy*. A page from the *Scientific Rochestarian* is reproduced below.]

RBP of R J Nash from *CHEMunications* April 1, 1979 p 6 (annotated by R J Nash).

The epic flight of Voyager II came to an abrupt end recently. Apparently, the mishap occurred while scientists at JPL were preoccupied with analysis of the data from the Jupiter fly-past.

Because of cut-backs in federal funds, the JPL controllers had been furloughed and told to report back in July 1985 in time for the Uranus fly by. Consequently, mission-control was staffed with only a skeleton crew of Kelly Girls, and a programmer, in a fit of boredom, updated the on-board computer using a copy of 8K BASIC Star Trek. In response, the guidance slewed from Canopus and locked-on to Antares, and Voyager II's course changed from a close encounter to a direct hit on Uranus.

As JPL officials watched helplessly, Voyager II raced towards Uranus and, following one last transmission of: -.-. .-. ..- -. -.-. ...., all signals ceased. Using long-range cameras aboard Voyager III (launched two months after Voyager II), scientists have located the impact site, and the following is a computer reconstructed picture.

Interviewed on the Johnny Carson show, Carlos Sago, Chairman of Astralchemistry at Cornwell University,

stated, 'Clearly the current theories are wrong! I intend to apply to NASA for funding of an eggshell model of planetary crust formation'.

## 150 YEARS AGO

'Using plates coated with bitumen and lavender oil, J N Niepce has determined to create permanent pictorial records. However, Niepce's plates, following exposure to strong sunlight, reveal only a random pattern of light and dark areas. To forestall financial ruin, Niepce now intends to offer his exposed plates as perfumed floor-tiles.'

## 70 YEARS AGO

'The future of the newly formed Research Laboratory at the Eastman Kodak Company has been seriously compromised by the abrupt return of Dr C E K Mees to England. It appears that Dr Mees was striken with a debilitating attack of homesickness immediately upon his arrival in Rochester, NY. At last report, Mees was making a miraculous recovery at the Croydon Cricket Club. Meanwhile, the entire stock of Wratten & Wainwright plates has been sold to Messrs. Lord & Burnham, who intend to enter the greenhouse business.'†

## 50 YEARS AGO

'The highlight of the current music season must surely be the recent performance of Mozart's F-Major Quartet, K590, by violinists Leopold D Mannes and Leo Godowsky Jr and members of the Eastman School of Music. George Eastman will long be remembered for his steadfast patronage of Mannes and Godowsky—this pair entertained thoughts of photographic research, but Eastman wisely dissuaded them from such activities, noting that music would last longer than any color film.'‡

†George Eastman actually soled Dr Mees's reluctance to leave England and the Wratten & Wainwright company, by buying the latter company!
‡Even though Godowsky and Mannes were indeed professional musicians, they did (as keen amateur photographers) lay the foundations for Kodachrome color film (so much for the Kodak Research Labs!).

426

## 30 YEARS AGO

'C Carlson, lately a patent attorney, dreams of a process which would generate unlimited copies, and, despite rebuffs from the finest technological companies in America, he continues to search for others who would share his conviction. His closest success to date has been with the Haloid Corp. Apparently, the Director of Research at Haloid chanced upon a description of Carlson's process, and brashly proposed that Haloid develop Carlson's ideas. This notion was firmly rejected by the Comptroller, who noted that Haloid's future lay in a growth industry. Rumors circulate that Haloid has taken an option on 1000 acres of apple orchard in Webster, NY, and intends to become a major producer of apple-sauce.'†

†The Haloid Corporation was a minor producer of photographic paper, before gambling on the new copying process invented by Chester Carlson. From such a humble beginning came the present Xerox Corporation, which is indeed sited in Webster, NY, on land formerly used as an apple orchard. (The city of Rochester thought that one monolithic Kodak Park was sufficient, and accordingly rebuffed the Xerox Corporation when the latter company began its explosive growth in the 60's. Thus, the Xerox Corporation chose tax-hungry rural Webster for its expanded R&D and Manufacturing plant.)

---

# The demise of the slide rule

*K M Reese*

RBP from *Chemical & Engineering News* **60**, 18 Jan 1982. © 1982 American Chemical Society.

*The New York Times* for January 3 carried a story on a vanishing species—the slide rule. Reporter Kirk Johnson wrote that in 1967, Keuffel & Esser of Hoboken, NJ, then the leading U.S. maker of slide rules, was commissioned to do a study of the future. The study forecast a number of technological developments, but not the demise of K&E's best-known product. 'Nobody really foresaw the cheap calculator', the company's John Montesi told reporter Johnson.

The last U.S. maker of slide rules probably was Sterling Plastics of Mountainside, NJ, according to the *Times*. Sterling stopped production in December 1980. Its leading product in the few years before the end was a $1.79 plastic slide rule, says the company's John Heath.

K&E made the Cadillac of slide rules—a mahogany device about a foot long. It came in a dandy leather scabbard that you could hang on your belt and it cost about $40. Sales peaked in the late 1950s, when the company sold about 20,000 slide rules a month. K&E stopped production in the early 1970s, routed by the cheap calculator. It still has about 2300 slide rules in stock and sells about 200 of them a year.

Both K&E and Sterling evidently survived the slide-rule debacle in good shape. In fact, two thirds of K&E's stock was bought late in 1981 by Kratos Inc., La Jolla, Calif., a maker of analytical devices, aircraft instruments, and computer display equipment, the *Times* says.

There was a day when no self-respecting engineering student would be caught dead without a foot-long holstered slide rule slung from his belt, but no more. This department has a 6-inch K&E said to be more than 40 years old. It still works perfectly.

## Errors and myopia

From W J Rees 1901 *The Smithsonian Institution; Documents Relative to its Origin and History*. (Washington: US Government Printing Office) p 617.

During debate in the Senate (February 21, 1861) on an appropriation for the Smithsonian Institution, Senator John P Hale made the following criticism:

'... I have devoted some time every year, more or less, to finding out what on earth that Smithsonian Institution was for; I have had friends who have visited Washington, who have told me that they were going to examine it to find out; and I have asked them repeatedly, if any of them had found it out, to tell me. The New York Tribune ... said that it was a sort of lying-in hospital for literary valetudinarians. ... I am opposed to the whole of it, I think it is wrong.'

\*\*\*\*\*\*

Rear Admiral
G W Melville
*North American
Review* December
1901.

Rear Admiral George W Melville wrote:

'... there is no basis for the ardent hopes and positive statements made as to the safe and successful use of the dirigible balloon or flying machine, or both, for commercial transportation or as weapons of war, and that, therefore, it would be a wrong, whether wilful or unknowing, to lead the people and perhaps governments at this time to believe the contrary ...'

\*\*\*\*\*\*

T A Edison *North
American Review*
November 1889.

Thomas Edison preferred DC networks for electric power and fought proposals for using alternating current. He wrote, in 1889:

'There is no plea which will justify the use of high-tension and alternating currents, either in a scientific or a commercial sense. They are employed solely to reduce investment in copper wire and real estate ...
'... My personal desire would be to prohibit entirely the use of alternating curents. They are unnecessary as they are dangerous... I can therefore see no justification for the introduction of a system which has no element of permanency and every element of danger to life and property.'

\*\*\*\*\*\*

Sir Ernest Rutherford, director of the Cavendish Laboratory and father of nuclear physics, in a speech for the British Association for the Advancement of Science in 1933 gave this opinion:

'We cannot control atomic energy to an extent which would be of any value commercially, and I believe we are not likely ever to be able to do so.'

# 4004 B.C.

[James Ussher (1581 – 1656), Archbishop of Armagh and Primate of All Ireland, was highly regarded in his day as a churchman and as a scholar. Of his many works, his treatise on chronology has proved the most durable. Based on an intricate correlation of Middle Eastern and Mediterranean histories and Holy Writ, it was incorporated into an authorized version of the Bible printed in 1701, and thus came to be regarded with almost as much unquestioning reverence as the Bible itself. Having established the first day of creation as Sunday 23 October 4004 B.C., by the arguments set forth in the passage below, Ussher calculated the dates of other biblical events, concluding, for example, that Adam and Eve were driven from Paradise on Monday 10 November 4004 B.C., and that the ark touched down on Mt Ararat on 5 May 1491 B.C. 'on a Wednesday'.]

James Ussher 1658 *The Annals of the World ix*, quoted by G Y Craig and E J Jones in *A Geological Miscellany* (Oxford: Orbital Press) 1982.

For as much as our Christian epoch falls many ages after the beginning of the world, and the number of years before that backward is not onely more troublesome, but (unless greater care be taken) more lyable to errour; also it hath pleased our modern chronologers to adde to that generally received hypothesis (which asserted the Julian years, with their three cycles by a certain mathematical prolepsis, to have run down to the very beginning of the world) an artificial epoch, framed out of three cycles multiplied in themselves; for the Solar Cicle being multiplied by the Lunar, or the number of 28 by 19, produces the Great Paschal Cycle of 532 years, and that again multiplied by fifteen, the number of the indiction, there arises the period of 7980 years, which was first (if I mistake not) observed by Robert Lotharing, Bishop of Hereford, in our island of Britain, and 500 years after by Joseph Scaliger fitted for chronological uses, and called by the name of the Julian Period, because it contained a cycle of so many Julian years. Now if the series of the three minor cicles be from this present year extended backward unto precedent times, the 4713 years before the beginning of our Christian account will be found to be that year into which the first year of the indiction, the first of the Lunar Cicle, and the first of the Solar will fall. Having placed therefore the heads of this period in the kalends of January in that pro-

430

leptick year, the first of our Christian vulgar account must be reckoned the 4714 of the Julian Period, which, being divided by 15. 19. 28. will present us with the 4 Roman indiction, the 2 Lunar Cycle, and the 10 Solar, which are the principal characters of that year.

We find moreover that the year of our fore-fathers, and the years of the ancient Egyptians and Hebrews were of the same quantity with the Julian, consisting of twelve equal moneths, every of them conteining 30 days, (for it cannot be proved that the Hebrews did use lunary moneths before the Babylonian Captivity) adjoying to the end of the twelfth moneth, the addition of five dayes, and every four year six. And I have observed by the continued succession of these years, as they are delivered in holy writ, that the end of the great Nebuchadnezars and the beginning of Evilmerodachs (his sons) reign, fell out in the 3442 year of the world, but by collation of Chaldean history and the astronomical cannon, it fell out in the 186 year of Nabonasar, and, as by certain connexion, it must follow in the 562 year before the Christian account, and of the Julian Period, the 4152, and from thence I gathered the creation of the world did fall out upon the 710 year of the Julian Period, by placing its beginning in autumn: but for as much as the first day of the world began with the evening of the first day of the week, I have observed that the Sunday, which in the year 710 aforesaid came nearest the Autumnal Æquinox, by astronomical tables (notwithstanding the stay of the sun in the dayes of Joshua, and the going back of it in the dayes of Ezekiah) happened upon the 23 day of the Julian October; from thence concluded that from the evening preceding the first day of the Julian year, both the first day of the creation and the first motion of time are to be deduced.

# Portrait of the scientist as a late victorian ruin

*Ted Ward*

Not for nothing was Dr Flay called the Don Bradman of Chemistry. Like Bradman he had 'batted' on many a sticky wicket but, relentless in acquiring power and glory, he already had over two hundred chemical research papers to his name. Having achieved the double century many might have thought it time to ease off a bit, especially as Flay was nearing sixty and only a few more years to go before he retired from the Headship of the Chemistry Department at Blandfield Technical College. However, like Bradman, the more 'runs' he made the more he wanted to make. He now boasted openly that he would get three hundred papers out before he retired. Callers to his office would get the chance of seeing what his achievements were all about. Flay would take from his desk a rather dirty looking stiff covered exercise book and opening it would disclose, in his rather erratic almost unreadable writing, all the details of his multitudinous collection of published papers. Unlike Don Bradman, however, Flay really only had quantity to offer, whereas Bradman invariably combined quantity with classical quality. For real lack of quality Flay often found it difficult to get his stuff into print in the best scientific journals. But Flay had devised ways of overcoming this obstacle. He had a number of second rate, even third rate, journals in tow, who would be only too glad to fill up their pages with Flay's latest touch of genius (of course we are talking of the nineteen forties here, it was much easier to get papers into print in those days, especially in lower grade journals). Just to make absolutely certain of getting everything into print (and he never seemed to fail whatever the subject or his treatment of it, even an article on *World Peace Through Research*! got into an obscure textile journal!) he sat on the publications board of one of these less reputable journals. His attendances at their meetings were rare and largely determined by the need for Flay to get something in print, rather than to serve the needs of others. Even during the Second World War, when paper for scientific journals was very hard to come by indeed, his output did not diminish. In fact it increased since he had fewer rivals, the majority of his fellow chemists being engaged on war work up to their necks. Flay had decided he had done enough in World War I of that sort of thing and refused to help in this new war, apart from an odd bit of fire watching at the college. Of course this behaviour earned him many a black mark from his scientific contemporaries but Flay himself saw nothing wrong in his attitudes. One journal editor had humorously suggested at an academic meeting that he, Flay, had used up most of the journal's paper ration, a remark that had a real grain of seriousness about it. Flay, however, found this situation very amusing and relayed the story to everyone who would listen.

It was quite clear that Flay's enthusiasm for chemistry, which none who knew him disputed, overcame the scientist's need to proceed cautiously in making claims for his work. Indeed Flay had arrived at the point where he practically had no standards at all. His enthusiasm for chemistry was only exceeded by his enthusiasms for himself and for money (in various forms). The thing he liked talking about best was himself, his many achievements and experiences. At the other end of the scale he was an extremely bad listener, just anxiously going along with what was being said before he could put his spoke in again. However, since he had a robust Lancashire manner about him, combined with a really gusty sense of humour (often collapsing into uncontrollable laughter at his own jokes), this relentless egomania was more tolerable, some even found it irresistible. It was this gusty humour that made him a very popular lecturer at gatherings of chemists. To the scientist who has to sit through many lectures which are very exacting to follow, or understand, it was an immense relief to hear Flay 'on tour'. Yes, Flay on tour, because he liked to approach the business of lectures like a music hall booking, for the music hall was another of his favourite passions. Through some service to the conductor of the band he had sat in the orchestra pit at his local music hall, one night every week, with free admission. He had become a familiar figure over the many years he had gone there and was delighted if a famous music hall star drew attention to him as the conductor's uncle or some such crack. So 'on tour' his audience got a good ration of the typical Flay jokes and reminiscences, all mixed in with the chemistry, and they rarely failed to respond with hearty laughter, which delighted Flay, who would rush back from a tour of lectures to tell all his assistants about the triumph he had had at such and such a University. Nevertheless, like the actor he fundamentally was he had an extreme sensitivity and concealed this with his humorous outbursts. If one looked carefully enough one could detect the anxiety and ultimate lack of confidence in the engine room, probably brought on by his lowly origins in life.

Flay had once said that he could have become a chemist, a bishop or a comedian, one wondered if the role of bishop – comedian might not have suited him best. He might had stirred the Church up a bit and filled the pews to listen to his own special brand of religion. In fact on holidays, which he *always* spent at nearby Blackpool, he spent a lot of time reading the Bible or Paley's comparative religion.

Dr Flay did not choose to live in Blandfield, instead he preferred the relative isolation of the nearby town of Rumley, where in fact he had been born, which also seemed to show an uncertainty to move into the world at large, a need to remain in his childhood haunts. His house was on the edge of the slums, not very grand either but it enabled him to keep a watching eye on the many properties he owned in the slums themselves. He got a lot of satisfaction out of being a landlord on the large scale and it certainly gave him power over people's lives, which he craved for in many other directions. Knowing Flay's meaness the tenants probably found it very hard to get their repairs done but Flay was not without some friendly interest in his slum neighbours. One great story of his about this concerned some liver pills that were distributed as free samples in the area. These pills turned the urine green and much to Flay's astonishment he looked out of his bedroom window one morning to see two women showing each other pisspots over their garden walls full of green urine. Flay did a good deed then by going out of his way to tell the people not to worry, this was the usual effect of these pills and the incident became a standard joke on his lecture tours.

Flay did not like anyone to call at his house and when a research student tried this Flay was busy doing a spot of whitewashing (very appropriate to Flay!) and became very embarrassed. Flay made it clear to the intruder that he was not wanted and the student soon cleared off. Every morning, except in vacations (when he never visited the college at all, even though research students only had a months holiday), Flay journeyed to Blandfield by train, always aiming to get to the college before nine (if only to check up that his research students had arrived and that they were getting down to business). The railway staff knew him well, as also they knew Flay's little trick of walking up and down the platform to espy a compartment where someone had left a newspaper, so saving Flay the cost of buying one (*every* penny, but every penny, that Flay spent was carefully thought about, even though he had a good salary and received money from industry as a consultant). Flay, of course, was the sort of man whom people did remember, standing out as he did from the crowd. Over six feet tall, with a well made burly frame, marred only by an appreciably large sized pot belly that stuck out, ludicrously, in front, his figure topped with a massive balding red flushed head. He had run to fat in other places than his belly, he had dewlaps on his florid face, which held a rather bulbous large long nose, his most dominating characteristic. Looking carefully one noticed the beedy nervous dark eyes which swivelled about relentlessly. For such a large man he had rather small ears, which made him look like a character out of *Midsummer Nights Dream*. He usually wore a voluminous faded grey gaberdine raincoat (making him an even bigger figure on the landscape) and beneath this a blue serge suit. Not ordinary blue serge but massively thick stuff that one would have thought could not be bought in 1940, if not many years before that. Rumour had it that the suit had been passed on from his grandfather to his father and then to Flay. Just to guard against any possibility that his appearance would have to be modified in any way (he seemed determined, absolutely determined, except in his organic chemistry, to live in the past) he had bought up the last remaining bolt of this ironclad blue serge that existed in Rumley. If his old faded suit gave out, then he could have another carved out from his stock. On his massive head he sported a very large black bowler of fascinating construction. It had a curly brim and was butressed by black lace ties at the side. One of his younger members of staff claimed that he had researched into such hats and declared that nothing like it had been made since 1906. So Flay's hat, like Flay himself, was a veteran indeed. Yes, that just about summed up Flay (if one can sum up an infinitely complex character like Flay), a veteran of long forgotten campaigns, a veteran of out-moded customs and manners, a veteran of ancient ideas but not in his Chemistry. There his views seemed to be so far advanced that many distinguished chemists declared them to be incomprehensible. But more of Flay's chemical views later, we must now delve deeper and find out what lay behind his outer garments, delving which revealed and emphasised that Flay was a super miser.

He always wore a black openwork silk tie that had been given him by a research student way back in 1926. As for his shirts a full archeological investigation was required to determine their origins. These amazing shirts were pieced together from portions of old shirts (or had been repaired that way from time to time), one student had counted thir-teen different patterns of material in the visible portion of one shirt, with possibly others in the unseen areas! Could anyone be meaner than that and yet at the same time boast of a fortune which he hinted at came to a hundred thou-sand pounds or so. Flay had no inhibitions about his dress or his mannerisms. Showing his aggressive spirit he would firmly hold his fading umbrella at right angles to his body, a lovely way to cleave through a crowd when you are a big man like Flay. In the other hand he had a further weapon, a massive battered Gladstone bag, not for Flay anything as modern as a smart looking briefcase.

As he thrust his way through the crowds his mind was, more often than not, thinking about his latest theories. His head was filled with whirling atoms, molecules and electrons, especially electrons, to Flay the real stuff of chemistry. These various particles were all plunging wildly all over the place but Flay was determined to pin them down. He literally seized them out of the air and set them down on paper with lovely symbols, and then festooned them about with arrows (curly and otherwise) to denote their movements, adding a black dot here and there for his beloved electrons. Flay then claimed that these were the directions in which the particles moved in his reaction schemes and God help anybody who thought otherwise. Yes, it was an exciting, feverish, kind of world that Flay lived in, his central nervous system constantly agitated by all these whirling thoughts (which he could hardly control at times), continual stimulation of the body juices via vibrant electrons. Nevertheless it was these wild schemes for describing how atoms and electrons got from one arrangement to another that was Flay's real weakness. But his enthusiasm was infectious, his humour unstoppable and fruity, but as the editor of one journal put it 'its not your practical work Flay that gets me down, its your bloody theory'.

Another aspect of Flay's world which provided further stimulation and excitement (if something was not that way for Flay he soon had to make it that way, life for this actor would otherwise be unbearable) were his scientific opponents, his *enemies* as he believed. He seemed to be surrounded by them at every turn, aided and abetted, of course, by his college colleagues who always seemed to be on the opposite side to Flay (small wonder as Flay saw very little of other staff and kept himself to himself in the college, rarely venturing out of his department). These enemies included editors of journals, referees of chemical papers (as these assessors are called in the chemical world), professors in this and that University, chemists in large industrial firms, etc. These were the people that were preventing Flay from getting his due recognition in life, this was the reason, so Flay believed, why he was stuck away in a back street at Blandfield at a second rate technical college, instead of being Lord and Master of some great University department. What a humiliation for Flay, once a brilliant scholar at an ancient University, mixing with the likes of Eddington, men who would change man's concept of the physical world, who would achieve international fame and receive Nobel prizes. Flay had obtained a Ph.D. at one of Germany's most famed and ancient Universities. Yet Flay was destined to stay in this smoke ridden, chemical ridden, slum ridden backwater (all true of Blandfield in those days, only relieved by the amazing prowess of the town's soccer and rugby league teams, an analgesic for the town's downtrodden citizens), hacking away all day at his enemies, inside and outside the college. Oh yes he had his enemies all right, even in the college itself, in fact almost the whole college seemed to be actively conspiring to prevent Flay getting his due reward (an F.R.S. and maybe a few honorary doctorates from various universities). Later on, to his delight and amazement, he was nominated for election to the Royal Society, one of the very few that had been so nominated from a technical college. He had had his eye on an F.R.S. for decades and here were six extremely important professors of chemistry, probably the six best known figures in chemistry in Great Britain, nominating him for election. Naturally Flay thought it was in the bag with sponsors like these but, to his chagrin, he visited an F.R.S. friend in Scotland (Flay had plenty of F.R.S. friends, they loved his jokes even if they didn't like his theories!), who showed him confidential reports on candidates that were being circulated to members of the Royal (prior to each year's elections) and it revealed that these same six sponsors had nominated at least a dozen other chemists, also that some peoples' nominations went back decades and yet had never been elected to one of the most eminent bodies of scientists in the world (it was not unknown for people to get in by reason of their social standing and service on committees of all kinds, rather than the quality of their science, one duffer had been elected simply because he had married the sister of a President's wife).

For a time after this incident the Royal was included in Flay's enemies but since he had five years to run with his candidature, before renomination, he calmed down and even tried on one occasion to smooth down an F.R.S. who had done Flay a real injustice, and who normally would have been the subject of a savage polemic from Flay (he was a master at the art of polemics, no doubt encouraged by his German training as the Germans loved these vitriolic type of confrontations in print). This particular F.R.S. had carried out some very simple chemical experiments very badly indeed and then even more disastrously had used the worthless results as the basis for a lot of speculative theory. It was a golden opportunity for Flay to chew his opponent into small pieces and spit the pips out! God knows how he restrained himself but restrain himself he did. He got one of his assistants to repeat the experiments properly, then he wrote a very diplomatic research paper in which he, ever so gently, tore a strip off from his victim. Even when the F.R.S., in a rejoinder, criminally side-stepped the issues and obscured the whole business by quoting obscure foreign work, Flay just let the whole thing slide, despicable as his opponents tactics had been. For once Flay was definitely on the side of the angels !

Anyway, he, Flay, would show them, yes he would indeed show his enemies that the back street technical college would not prevent Flay 'coming to the front' as he quaintly put it. Flay had fantasies about swigging old port in college combination rooms with eminent Dons, whilst his erstwhile down-trodden technical college contemporaries were still trying to ram knowledge into unwilling brains that had been busy all day for their employers and were now struggling to higher things at evening classes. However, in Blandfield Flay was free of all this routine lark; admin and teaching he left almost entirely to his staff and further humiliated them by giving his senior laboratory steward authority over them in many aspects of their lives. This steward had once been a wool scourer but was a highly skilled practical man. He was utterly devoted to 'the doctor' as he liked to describe his boss. One day the steward excitedly burst into Flay's office with the news that Professor Crosson had died the previous night, Crosson being a close rival of Flay from a nearby University, who worked in similar fields to Flay but with much less success. Crosson had got into the Royal because he had married a woman who had two close relatives who had become very eminent members of that body. It is appalling to record that Flay and his steward were literally rubbing their hands at this sad news, seeing it as a real chance for Flay to pop into the vacancy. It did not stop Flay writing a fullsome obituary of Crosson, praising Crosson to the skies (sad to record that when Flay died the same journal printed a tiny obituary on him written by a very minor figure. Flay must have turned in his grave at this final victory for his enemies).

Flay confined himself for practically the whole day to his office and private research laboratory. Now and again, when some new enthusiasm struck him (which was all too often with Flay's whirling type of brain) he would dash out to tell his research students of his latest brilliant scheme and of all the enemies who would conspire to stop him getting it into print, and of all the enemies who would condemn it when it was in print. Even the gents toilet was not a sacred place, one of his men might be getting down to a quiet shit when Flay would come along and his loud enthusiastic voice would come over the top of the cubicle. The poor unwilling victim would then be engaged in chemical conversation, rounded off by the usual fruity jokes (such as the one about the man who went to a Salvation army meeting and was confronted by a Sally demanding to know if he had been saved. 'No' said the

440

man 'I have not, do you save women as well?'. 'Yes' ardently replied the Sally, to which came the rejoinder—'then save me a couple for next week') which sent Flay away in gusts of belly splitting laughter. He would be rapidly on his way to do a bit more stirring up in his room or to carry out a few experiments, in his nerve-wracked incompetent fashion, in his private laboratory. If students ventured into the inner sanctum they saw an amazing sight. Flay in his undress uniform. He never worked in a lab coat, always in his waistcoat and shirt (of many portions and patterns, as we have remarked before). His shirt sleeves would be rolled up and poking below them a few inches of thick wooly vest. Of course he would be sporting his black silk tie, no one ever saw him wearing any other tie than this. Students were very rarely admitted to Flay's private laboratory and if they somehow ventured in were rapidly ushered out again. The reason for this was that Flay kept enormous stocks of valuable chemicals in his lab, stuff he, in his meaness, did not intend to give to his students, even though obviously it would have speeded up their work and helped Flay's productivity of papers (for that's what Flay was now, a research paper producing machine, numbers alone counted and backed up by his 'music hall' lecture tours that's what Flay thought would get him the F.R.S., although only two scientists from technical colleges had ever got the coveted honour). Another type of confrontation with students came if they disturbed him in his lunch break (Flay made no attempt to join other staff in the refectory, not even his own departmental staff). What a disgusting sight confronted the lunch time intruder. Flay munching away at his sandwiches (which his wife had put up in his battered sandwich tin), often in his shirt sleeves with vest showing, and swilling it all down with tea prepared in a dirty tea stained beaker. Or later in his lunch hour they would find him grubbing a few more pennies, abstracting items from journals, for which he would be paid at a very low rate and wasting valuable research time.

Although his students liked his broad humour and other

antics they did not think much of him as a research supervisor. They would tolerate him demonstrating, pot belly and all, a Ginger Rogers dance routine after he had been stimulated by a visit to the cinema (where he and his wife sat through *all* the performances, having brought sandwiches to succour them). They would tolerate him telling a succession of jokes from the music hall or bursting into song with arias from grand opera (relics of his student days in Germany). But when they tried to pin him down to definite statements about the work in hand he always seemed to wriggle away from them and usually leave them as bewildered as when they first questioned him. No wonder that he was universally known as the 'old bugger' and, since they were under his thumb for money, they had to take it (unlike most establishments where research students get their scholarship money at the beginning of term Flay had arranged to pay it in arrears with the constant threat that if they did not toe the line they might not get it at all). Then at the end of three years they had to deliver up their theses which Flay would then cut up into as many pieces as possible for publication.

There was one thing the students never had to worry about, the analyses of their products. Flay and his ex-wool scourer steward did all these and everything that went through the system seemed to emerge with universal approval. Where the analysis depended on a measurement of gas Flay would have a shrewd idea how much gas he wanted to get into the measuring vessel and literally cut the apparatus off when he had the correct volume. There was a very embarrassing situation one day. An old research student called in for a chat and Flay was enthusiastically demonstrating his powers as an analyst. Sure enough he got a marvellous result but the visitor had spotted that Flay had heated the combustion tube so strongly that it had developed a hole, letting in nitrogen from the air and so ruining his nitrogen analysis. Flay was so flustered and bewildered that for once his aplomb and humorous rejoinders deserted him, he did not know where to put himself and the visitor, of course, earnt a

442

monumental black mark in Flay's enemies book, it was as if the unfortunate chap had carved the hole in the tube himself. After that, 100% secrecy was vested in the analysis operations. Constant complaints came in from other chemists about results Flay had published and which they could not repeat (usually by private letter as most chemists were decent people and not given to unnecessary public exposure without fully justified cause). Flay, in contrast to his relentless aggressive attitude to his research rivals, just shrugged off the complaints and got on with the next job in hand.

Research students had much to complain about their basic raw materials. These were often relatively crude chemicals shipped in by the barrel from Mammoth Chemicals Unlimited, who paid Flay a retaining fee for his advice from time to time. Old students would exchange cracks about what barrel their Ph.D. had emerged from!

Flay had another unforgivable sin to which we have already referred, he was about the meanest man on God's earth. Although he was always boasting about the money he had and the people he could buy up with it (his enemies) he was careful, utterly careful, to spend as little as possible. His ingenious brain concocted great schemes for saving money at all times. He proudly told the story of how he was once called up for an interview for a job in London. He managed to buy two excursion tickets (that joined up with one another) to get him there and even took in a rugby match in Leicester at the break point in his journey. When he got himself elected to a professional committee in London, later in his life, he worked out a scheme whereby he rose extremely early, got a succession of workmen's tickets for about half the journey, finished off on a third class ticket and claimed first class travel expenses. If he stayed overnight he did so with relatives and again claimed first class hotel expenses. His mind was sharply tuned in to such matters. Once a student told him he had to go to Rumley the next day. In a flash Flay declared that he must go by bus as he would save a halfpenny! His greatest triumph came one day when fog prevented his train stopping at his usual local station in Rumley and he was carried on to the terminus. The extra charge would only have been a halfpenny, indeed if any had been demanded, but Flay was absolutely exultant when he swept through the barrier declaiming 'you can't sue me for a halfpenny in law'.

The kind of chemistry that Flay went in for threw up some further interesting points about his character. Just as dog owners choose dogs that often resemble them in appearance or character, or biologists work with animals that bear a similar relationship, chemists often work with materials that suit their personalities. A colourful person with an explosive approach to chemistry, a man who under his tough Lancashire armour was incredibly sensitive to everything around him, it was not surprising that Flay worked with the theons. Not only did they come in a variety of exotic colours but their reactions went extremely fast, giving off masses of gas. These theons suited his nervous, impatient nature, everything was soon over and done with, no staring at flasks boiling on tripods for hours, no great demands on technique (some students thought that Flay had no technique but fudged everything and they might well have been right), also it was all very exciting and dramatic.

The nearer he drew to his retirement the more acute became Flay's sufferings. He saw the coveted F.R.S. slipping away from him. His new eccentricity appeared. Fantastic explanations about why he had not succeeded, starting with a terrible story about an eminent chemist who had been killed mountaineering in the Alps on his honeymoon. Flay, being at that time under the deceased's wing, was held up in his march forward. No sympathy for the dead man, just that he had been incredibly wrong to get himself killed like that. Another beautiful story was about when he applied for a professorship in Scotland and the man appointed was chosen for his knowledge of buildings (they proposed to put up a new building for him) and his ability in research. To Flay's chagrin the gentleman in question did no research whatsoever and during his tenure of office no new construction was started.

As the end of his time approached he repeatedly stated that since 'he had come in with bare boards' that was the way he intended to leave, he cleared everything out that might have been of use to his successor and gave that man no help in establishing himself. As soon as he retired he cut himself off from all chemists, set himself up in a large seaside villa (despite the expense involved, the agony must have damned nigh killed him) and probably spent the rest of his life explaining to himself why he had not succeeded. One of his members of staff summed him up in an utterly classical chemical joke. Flay was saying, reflectfully to him in a pre-retirement discussion 'that all my electronic shells are filled now' (a state that removes reactivity in chemical compounds). Quick as a flash came the rejoinder 'Yes, you've become an inert gas'.

# For further reading

Leads into the literature of humor in science may be found in:

R A Baker 1963 (ed) *A Stress Analysis of a Strapless Evening Gown* (Prentice-Hall) 192 pp

Stephen Castell 1983 *Computer Bluff* (Quartermaine House Ltd) 120 pp

Norman E Dodson 1981 *Math Poetry and Stuff* (Carlton Press) 95 pp

D S Halacy Jr 1967 *Science and Serendipity, Great Discoveries by Accident* (Macrae Smith Co) 155 pp

Alexander Kohn *Humor, the Interdisciplinary Denominator in Science: Interdisciplinary Science Reviews* vol 7 no 4 pp 309–324

Ralph A Lewin 1981 *The Biology of Algae* (University Press of America) 103 pp

Alan L Mackay 1976 *The Harvest of a Quiet Eye* (IOP Publishing) 204 pp

Richard F Mould 1984 *Mould's Medical Anecdotes* (IOP Publishing) 160 pp

Ralph Paddington 1963 *The Psychology of Laughter, A Study in Social Adaptation* (NY: Gamut)

John Read 1947 *Humor and Humanism in Chemistry* (London) 388 pp

C S Slichter 1940 *Science in a Tavern, Essays and Diversions on Science in the Making* (University of Wisconsin Press) 206 pp

# Index

# Index